SCHAUM'S
EASY OUTLINES

Statistics

—————— *Second Edition*

Murray R. Spiegel

Abridgement Editor:
David P. Lindstrom

New York Chicago San Francisco Lisbon London Madrid Mexico City
Milan New Delhi San Juan Seoul Singapore Sydney Toronto

The McGraw·Hill Companies

Copyright © 2011 by The McGraw-Hill Companies, Inc. All rights reserved. Printed in the United States of America. Except as permitted under the United States Copyright Act of 1976, no part of this publication may be reproduced or distributed in any form or by any means, or stored in a database or retrieval system, without the prior written permission of the publisher.

3 4 5 6 7 8 9 10 11 12 13 14 15 DIG/DIG 10 11 12

ISBN 978-0-07-174581-9
MHID   0-07-174581-5

Library of Congress Cataloging-in-Publication Data

Trademarks: McGraw-Hill, the McGraw-Hill Publishing logo, Schaum's, and related trade dress are trademarks or registered trademarks of The McGraw-Hill Companies and/ or its affiliates in the United States and other countries and may not be used without written permission. All other trademarks are the property of their respective owners. The McGraw-Hill Companies is not associated with any product or vendor mentioned in this book.

McGraw-Hill books are available at special quantity discounts to use as premiums and sales promotions or for use in corporate training programs. To contact a representative, please e-mail us at bulksales@mcgraw-hill.com.

This book is printed on acid-free paper.

Contents

Chapter 1
VARIABLES AND GRAPHS

IN THIS CHAPTER:

✔ *Inferential and Descriptive Statistics*
✔ *Variables: Discrete and Continuous*
✔ *Scientific Notation*
✔ *Functions*
✔ *Rectangular Coordinates and Graphs*
✔ *Raw Data, Arrays, and Frequency Distributions*
✔ *Histograms and Frequency Polygons*
✔ *Relative-Frequency Distributions*

Inferential and Descriptive Statistics

Statistics is concerned with scientific methods for collecting, organizing, summarizing, presenting, and analyzing data as well as with drawing valid conclusions and making reasonable decisions on the basis of such analysis. In a narrower sense, the term *statistics* is used to denote the data themselves or numbers derived from the data, such as averages. Thus we speak of employment statistics, accident statistics, etc.

In collecting data concerning the characteristics of a group of individuals or objects, such as the heights and weights of students in a university or the numbers of defective and nondefective bolts produced in a factory on a given day, it is often impossible or impractical to observe the entire group, especially if it is large. Instead of examining the entire group, called the *population*, or *universe*, one examines a small part of the group, called a *sample*.

A population can be *finite* or *infinite*. For example, the population consisting of all bolts produced in a factory on a given day is finite, whereas the population consisting of all possible outcomes (heads, tails) in successive tosses of a coin is infinite.

If a sample is representative of a population, important conclusions about the population can often be inferred from analysis of the sample. The phase of statistics dealing with conditions under which such inference is valid is called *inferential statistics*, or *statistical inference*. Because such inference cannot be absolutely certain, the language of *probability* is often used in stating conclusions.

The phase of statistics that seeks only to describe and analyze a given group without drawing any conclusions or inferences about a larger group is called *descriptive*, or *deductive*, statistics.

Before proceeding with the study of statistics, let us review some important mathematical concepts.

Variables: Discrete and Continuous

A *variable* is a symbol, such as X, Y, H, x, or B, that can assume any of a prescribed set of values, called the *domain* of the variable. If the variable can assume only one value, it is called a *constant*.

A variable that can theoretically assume any value between two given values is called a *continuous variable*; otherwise, it is called a *discrete variable*.

EXAMPLE 1.1. The number N of children in a family, which can assume any of the values 0, 1, 2, 3,... but cannot be 2.5 or 3.842, is a discrete variable.

EXAMPLE 1.2. The height H of an individual, which can be 62 inches (in), 63.8 in, or 65.8341 in, depending on the accuracy of measurement, is a continuous variable.

Data that can be described by a discrete or continuous variable are called *discrete data* or *continuous data*, respectively. The number of children in each of 1,000 families is an example of discrete data, while the heights of 100 university students is an example of continuous data.

 **Note!**

In general, *measurements* give rise to continuous data, while *enumerations*, or *countings*, give rise to discrete data.

It is sometimes convenient to extend the concept of variable to non-numerical entities; for example, color C in a rainbow is a variable that can take on the "values" red, orange, yellow, green, blue, indigo, and violet. It is generally possible to replace such variables by numerical quantities; for example, denote red by 1, orange by 2, etc.

Scientific Notation

When writing numbers, especially those involving many zeros before or after the decimal point, it is convenient to employ the scientific notation using powers of 10.

EXAMPLE 1.3. $10^1 = 10$, $10^3 = 10 \times 10 \times 10 = 1,000$.

EXAMPLE 1.4. $10^0 = 1$; $10^{-1} = .1$, and $10^{-5} = .00001$.

EXAMPLE 1.5. $864{,}000{,}000 = 8.64 \times 10^8$, and
$0.00003416 = 3.416 \times 10^{-5}$.

Note that multiplying a number by 10^8, for example, has the effect of moving the decimal point of the number eight places *to the right*. Multiplying a number by 10^{-6} has the effect of moving the decimal point of the number six places *to the left*.

We often write 0.1253 rather than .1253 to emphasize the fact that a number other than zero before the decimal point has not accidentally been omitted. However, the zero before the decimal point can be omitted in cases where no confusion can result, such as in tables.

Often we use parentheses or dots to show the multiplication of two or more numbers. Thus $(5)(3) = 5 \cdot 3 = 5 \times 3 = 15$, and $(10)(10)(10) = 10 \cdot 10 \cdot 10 = 10 \times 10 \times 10 = 1000$. When letters are used to represent numbers, the parentheses or dots are often omitted; for example, $ab = (a)(b) = a \cdot b = a \times b$.

The scientific notation is often useful in computation, especially in locating decimal points. Use is then made of the rules

$$(10^p)(10^q) = 10^{p+q} \qquad 10^p / 10^q = 10^{p-q}$$

where p and q are any numbers.

In 10^p, p is called the *exponent* and 10 is called the *base*.

EXAMPLE 1.6. $(10^3)(10^2) = 1000 \times 100 = 100{,}000 = 10^5$, i.e., 10^{3+2}
$10^6 / 10^4 = 1{,}000{,}000 / 10{,}000 = 100 = 10^2$, i.e., 10^{6-4}.

Functions

If to each value that a variable X can assume there corresponds one or more values of a variable Y, we say that Y is a *function* of X and write $Y = F(X)$ (read "Y equals F of X") to indicate this functional dependence. Other letters (G, ϕ, etc.) can be used instead of F.

The variable X is called the *independent variable*, and Y is called the *dependent variable*.

If only one value of Y corresponds to each value of X, we call Y a *single-valued function* of X; otherwise, it is called a *multiple-valued function* of X.

EXAMPLE 1.7. The total population P of the United States is a function of the time t, and we write $P = F(t)$.

EXAMPLE 1.8. The stretch S of a vertical spring is a function of the weight W placed on the end of the spring. In symbols, $S = G(W)$.

The functional dependence (or correspondence) between variables is often depicted in a table. However, it can also be indicated by an equation connecting the variables, such as $Y = 2X - 3$, from which Y can be determined corresponding to various values of X.

If $Y = F(X)$, it is customary to let $F(3)$ denote "the value of Y when $X = 3$," to let $F(10)$ denote "the value of Y when $X = 10$," etc. Thus if $Y = F(X) = X^2$, then $F(3) = 3^2 = 9$ is the value of Y when $X = 3$.

Rectangular Coordinates and Graphs

Consider two mutually perpendicular lines $X'OX$ and $Y'OY$, called the X and Y axes, respectively (see Fig. 1-1), on which appropriate scales are indicated. These lines divide the plane determined by them, called the XY plane, into four regions denoted by I, II, III, and IV and called the first, second, third, and fourth quadrants, respectively.

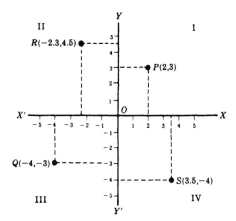

Figure 1-1

Point 0 is called the *origin*. Given any point P, drop perpendiculars to the X and Y axes from P. The values of X and Y at the points where the perpendiculars meet these axes are called the *rectangular coordinates*, or simply the *coordinates*, of P and are denoted by (X, Y). The coordinate X is sometimes called the *abscissa*, and Y is the *ordinate* of the point. In Fig. 1-1 the abscissa of point P is 2, the ordinate is 3, and the coordinates of P are $(2, 3)$.

Conversely, given the coordinates of a point, we can locate-or *plot* the point. Thus the points with coordinates $(-4, -3)$, $(-2.3, 4.5)$, and $(3.5, -4)$ are represented in Fig. 1-1 by Q, R, and S, respectively.

By constructing a Z *axis* through 0 and perpendicular to the XY plane, we can easily extend the above ideas. In such case the coordinates of a point P would be denoted by (X, Y, Z).

A *graph* is a pictorial presentation of the relationship between variables. Many types of graphs are employed in statistics, depending on the nature of the data involved and the purpose for which the graph is intended. Among these are *bar graphs*, *pie graphs*, *pictographs*, etc. These graphs are sometimes referred to as *charts* or *diagrams*. Thus we speak of bar charts, pie diagrams, etc.

Raw Data, Arrays, and Frequency Distributions

Raw data are collected data that have not been organized numerically. An example is the set of heights of 100 male students obtained from an alphabetical listing of university records.

An *array* is an arrangement of raw numerical data in ascending or descending order of magnitude. The difference between the largest and smallest numbers is called the *range* of the data. For example, if the largest height of 100 male students is 74 inches (in) and the smallest height is 60 in, the range is 74 - 60 = 14 in.

When summarizing large masses of raw data, it is often useful to distribute the data into *classes*, or *categories*, and to determine the number of individuals belonging to each class, called the *class frequency*. A tabular arrangement of data by classes together with the corresponding class frequencies is called a *frequency distribution*, or *frequency table*. Table 1-1 is a frequency distribution of heights (recorded to the nearest inch) of 100 male students at XYZ University.

Table 1-1

Height (in)	Number of students
60-62	5
63-65	18
66-68	42
69-71	27
72-74	8
Total	100

The first class (or category), for example, consists of heights from 60 to 62 in and is indicated by the range symbol 60-62. Since five students have heights belonging to this class, the corresponding class frequency is 5.

Data organized and summarized as in the above frequency distribution are often called *grouped data*. Although the grouping process generally destroys much of the original detail of the data, an important advantage is gained in the clear overall picture that is obtained and in the vital relationships that are thereby made evident.

A symbol defining a class, such as 60-62 in Table 1-1, is called a *class interval*. The end numbers, 60 and 62, are called *class limits*; the smaller number (60) is the *lower class limit*, and the larger number (62) is the *upper class limit*. The terms *class* and *class interval* are often used interchangeably, although the class interval is actually a symbol for the class.

You Need to Know

A class interval that, at least theoretically, has either no upper class limit or no lower class limit indicated is called an *open class interval*. For example, referring to age groups of individuals, the class interval "65 years and over" is an open class interval.

If heights are recorded to the nearest inch, the class interval 60-62 theoretically includes all measurements from 59 to 62 in. These numbers, which we can write as 59.5 and 62.5, are called *class boundaries*, or *true class limits*; the smaller number (59.5) is the *lower class boundary*, and the larger number (62.5) is the *upper class boundary*.

In practice, the class boundaries are obtained by adding the upper limit of one class interval to the lower limit of the next-higher class interval and dividing by 2.

Sometimes, class boundaries are used to symbolize classes. For example, the various classes in the first column of Table 1-1 could be indicated by 59.5-62.5, 62.5-65.5, etc. To avoid ambiguity in using such notation, class boundaries should not coincide with actual observations. Thus if an observation were 62.5, it would not be possible to decide whether it belonged to the class interval 59.5-62.5 or 62.5-65.5.

The size, or width, of a class interval is the difference between the lower and upper class boundaries and is also referred to as the *class width*, *class size*, or *class length*. If all class intervals of a frequency distribution have equal widths, this common width is denoted by c. In such case c is equal to the difference between two successive lower class limits or two successive upper class limits. For the data of Table 1-1, for example, the class interval is $c = 62.5 - 59.5 = 65.5 - 62.5 = 3$.

The *class mark* is the midpoint of the class interval and is obtained by adding the lower and upper class limits and dividing by 2. Thus the class mark of the interval 60-62 is $(60 + 62) / 2 = 61$. The class mark is also called the *class midpoint*.

For purposes of further mathematical analysis, all observations belonging to a given class interval are assumed to coincide with the class mark. Thus all heights in the class interval 60-62 in are considered to be 61 in.

Histograms and Frequency Polygons

Histograms and frequency polygons are two graphic representations of frequency distributions.

1. *A histogram, or frequency histogram*, consists of a set of rectangles having (*a*) bases on a horizontal axis (the *X* axis), with centers at the class marks and lengths equal to the class interval sizes, and (*b*)

areas proportional to the class frequencies.

If the class intervals all have equal size, the heights of the rectangles are proportional to the class frequencies, and it is then customary to take the heights numerically equal to the class frequencies. If the class intervals do not have equal size, these heights must be adjusted.

2. A *frequency polygon* is a line graph of the class frequency plotted against the class mark. It can be obtained by connecting the midpoints of the tops of the rectangles in the histogram.

The histogram and frequency polygon corresponding to the frequency distribution of heights in Table 1-1 are shown on the same set of axes in Fig. 1-2. It is customary to add the extensions PQ and RS to the next-lower and next-higher class marks, which have a corresponding class frequency of zero. In such case the sum of the areas of the rectangles in the histogram equals the total area bounded by the frequency polygon and the X axis.

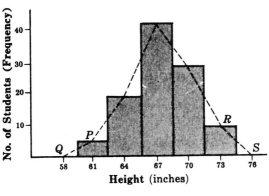

Figure 1-2

Relative-Frequency Distributions

The *relative frequency* of a class is the frequency of the class divided by the total frequency of all classes and is generally expressed as a percentage. For example, the relative frequency of the class 66-68 in Table

1-1 is 42 / 100 =42%. The sum of the relative frequencies of all classes is clearly 1, or 100%.

If the frequencies in Table 1-1 are replaced with the corresponding relative frequencies, the resulting table is called a *relative-frequency distribution*, *percentage distribution*, or *relative-frequency table*.

Graphic representation of relative-frequency distributions can be obtained from the histogram or frequency polygon simply by changing the vertical scale from frequency to relative frequency, keeping exactly the same diagram. The resulting graphs are called *relative-frequency histograms* (or *percentage histograms*) and *relative-frequency polygons* (or *percentage polygons*), respectively.

Chapter 2

MEASURES OF CENTRAL TENDENCY AND DISPERSION

IN THIS CHAPTER:

- ✔ Notation
- ✔ Averages, or Measures of Central Tendency
- ✔ The Arithmetic Mean
- ✔ The Weighted Arithmetic Mean
- ✔ The Median
- ✔ The Mode
- ✔ Quartiles, Deciles, and Percentiles
- ✔ Dispersion, or Variation
- ✔ The Standard Deviation
- ✔ The Variance
- ✔ Properties of the Standard Deviation

11

✔ *Absolute and Relative Dispersion, Coefficient of Variation*

✔ *Standardized Variable: Standard Scores*

Notation

Let the symbol X_j (read "X sub j") denote any of the N values $X_1, X_2, X_3,...,$ X_N assumed by a variable X. The letter j in X_j, which can stand for any of the numbers 1, 2, 3,..., N is called a *subscript*, or *index*. Clearly any letter other than j, such as i, k, p, q, or s, could have been used as well.

The symbol $\sum_{j=1}^{N}$ is used to denote the sum of all the X_j's from $j = 1$ to $j = N$; by definition,

$$\sum_{j=1}^{N} X_j = X_1 + X_2 + X_3 +...+ X_N$$

When no confusion can result, we often denote this sum simply by $\sum X$, $\sum X_j$, or $\sum_j X_j$. The symbol $\sum$ is the Greek capital letter *sigma*, denoting sum.

EXAMPLE 2.1. $\sum_{j=1}^{N} X_j Y_j = X_1 Y_1 + X_2 Y_2 + X_3 Y_3 +...+ X_N Y_N$

EXAMPLE 2.2. $\sum_{j=1}^{N} aX_j = aX_1 + aX_2 +...+ aX_N = a(X_1 + X_2 +...+ X_N)$ $= a\sum_{j=1}^{N} X_j$ where a is a constant. More simply, $\sum aX = a \sum X$.

EXAMPLE 2.3. If a, b, and c are any constants, then $\sum (aX + bY - cZ) = a \sum X + b \sum Y - c \sum Z$.

Averages, or Measures of Central Tendency

An *average* is a value that is typical, or representative, of a set of data. Since such typical values tend to lie centrally within a set of data arranged according to magnitude, averages are also called *measures of central tendency*.

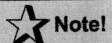 **Note!**

Several types of averages can be defined, the most common being the *arithmetic mean, the median,* and *the mode.* Each has advantages and disadvantages, depending on the data and the intended purpose.

The Arithmetic Mean

The *arithmetic mean,* or briefly the *mean,* of a set of N numbers $X_1, X_2, X_3, ..., X_N$ is denoted by $\bar{X}$ (read "X bar") and is defined as

$$\bar{X} = \frac{X_1 + X_2 + X_3 + \cdots + X_N}{N} = \frac{\sum\limits_{j=1}^{N} X_j}{N} = \frac{\sum X}{N} \qquad (1)$$

EXAMPLE 2.4. The arithmetic mean of the numbers 8, 3, 5, 12, and 10 is

$$\bar{X} = (8 + 3 + 5 + 12 + 10) / 5 = 38 / 5 = 7.6$$

If the numbers $X_1, X_2, ..., X_K$ occur $f_1, f_2, ..., f_K$ times, respectively (i.e., occur with frequencies $f_1, f_2, ..., f_K$), the arithmetic mean is

$$\bar{X} = \frac{f_1 X_1 + f_2 X_2 + \cdots + f_K X_K}{f_1 + f_2 + \cdots + f_K} = \frac{\sum\limits_{j=1}^{K} f_j X_j}{\sum\limits_{j=1}^{K} f_j} = \frac{\sum f X}{\sum f} = \frac{\sum f X}{N} \qquad (2)$$

where $N = \sum f$ is the *total frequency* (i.e., the total number of cases).

EXAMPLE 2.5. If 5, 8, 6, and 2 occur with frequencies 3, 2, 4, and 1, respectively, the arithmetic mean is

$$\bar{X} = [(3)(5)+(2)(8)+(4)(6)+(1)(2)] / (3+2+4+1) =$$
$(15+16+24+2) / 10 = 5.7$

The Weighted Arithmetic Mean

Sometimes we associate with the numbers $X_1, X_2,..., X_K$ certain *weighting factors* (or *weights*) $w_1, w_2,..., w_K$, depending on the significance or importance attached to the numbers. In this case,

$$\bar{X} = \frac{w_1X_1 + w_2X_2 + \cdots + w_KX_k}{w_1 + w_2 + \cdots + w_K} = \frac{\sum wX}{\sum w} \tag{3}$$

is called the *weighted arithmetic mean*. Note the similarity to equation (2), which can be considered a weighted arithmetic mean with weights $f_1, f_2,..., f_K$.

EXAMPLE 2.6. If a final examination in a course is weighted 3 times as much as a quiz and a student has a final examination grade of 85 and quiz grades of 70 and 90, the mean grade is

$$X = [(1)(70)+(1)(90)+(3)(85)]/(1+1+3) = 415 / 5 = 83$$

The Median

The *median* of a set of numbers arranged in order of magnitude (i.e., in an array) is either the middle value (if the number of data values is odd) or the arithmetic mean of the two middle values (if the number of data values is even).

EXAMPLE 2.7. The set of numbers 3, 4, 4, 5, 6, 8, 8, 8, and 10 has median 6.

EXAMPLE 2.8. The set of numbers 5, 5, 7, 9, 11, 12, 15, and 18 has median $\frac{(9 + 10)}{2} = 10$.

For grouped data, the median, obtained by interpolation, is given by

$$\text{Median} = L_1 + \left(\frac{\frac{N}{2} - (\sum f)_1}{f_{\text{median}}} \right) c \tag{4}$$

where L_1 = lower class boundary of the median class (i.e., the class containing the median)

N = number of items in the data (i.e., total frequency)

$(\sum f)_1$ = sum of frequencies of all classes lower than the median class

f_{median} = frequency of the median class

c = size of the median class interval.

Geometrically the median is the value of X (abscissa) corresponding to the vertical line which divides a histogram into two parts having equal areas.

The Mode

The *mode* of a set of numbers is that value which occurs with the greatest frequency; that is, it is the most common value. The mode may not exist, and even if it does exist it may not be unique.

EXAMPLE 2.9. The set 2, 2, 5, 7, 9, 9, 9, 10, 10, 11, 12, and 18 has mode 9.

EXAMPLE 2.10. The set 3, 5, 8, 10, 12, 15, and 16 has no mode.

EXAMPLE 2.11. The set 2, 3, 4, 4, 4, 5, 5, 7, 7, 7, and 9 has two modes, 4 and 7, and is called *bimodal*.

A distribution having only one mode is called *unimodal*.

In the case of grouped data where a frequency curve has been constructed to fit the data, the mode will be the value (or values) of X corresponding to the maximum point (or points) on the curve.

Quartiles, Deciles, and Percentiles

If a set of data is arranged in order of magnitude, the middle value (or arithmetic mean of the two middle values) that divides the set into two equal parts is the median. By extending this idea, we can think of those values which divide the set into four equal parts. These values, denoted by Q_1, Q_2, and Q_3, are called the first, second, and third *quartiles*, respectively, the value Q_2 being equal to the median.

Similarly, the values that divide the data into 10 equal parts are called deciles and are denoted by D_1, D_2,..., D_9 while the values dividing the data into 100 equal parts are called percentiles and are denoted by P_1, P_2,..., P_{99}. The fifth decile and the 50th percentile correspond to the median. The 25th and 75th percentiles correspond to the first and third quartiles, respectively.

You Need to Know

Collectively, quartiles, deciles, percentiles, and other values obtained by equal subdivisions of the data are called *quantiles*.

Dispersion, or Variation

The degree to which numerical data tend to spread about an average value is called the *dispersion*, or *variation*, of the data. Various measures of this dispersion (or variation) are available, the most common being the range, mean deviation, and standard deviation.

The *range* of a set of numbers is the difference between the largest and smallest numbers in the set.

EXAMPLE 2.12. The range of the set 2, 3, 3, 5, 5, 5, 8, 10, 12 is 12 - 2 = 10. Sometimes the range is given by simply quoting the smallest and largest numbers; in the above set, for instance, the range could be indicated as 2 to 12, or 2-12.

The Standard Deviation

The *standard deviation* of a set of N numbers $X_1, X_2,..., X_N$ is denoted by s and is defined by

$$s = \sqrt{\frac{\sum\limits_{j=1}^{N}(X_j - \bar{X})^2}{N}} = \sqrt{\frac{\sum(X - \bar{X})^2}{N}} = \sqrt{\frac{\sum x^2}{N}} = \sqrt{\overline{(X - \bar{X})^2}} \qquad (5)$$

where x represents the deviations of each of the numbers X_j from the mean . Thus s is the root mean square of the deviations from the mean, or, as it is sometimes called, the *root-mean-square deviation*.

If $X_1, X_2,..., X_K$ occur with frequencies $f_1, f_2,..., f_K$ respectively, the standard deviation can be written

$$s = \sqrt{\frac{\sum\limits_{j=1}^{K}f_j(X_j - \bar{X})^2}{N}} = \sqrt{\frac{\sum f(X - \bar{X})^2}{N}} = \sqrt{\frac{\sum fx^2}{N}} = \sqrt{\overline{(X - \bar{X})^2}} \qquad (6)$$

where $N = \sum_{j=1}^{K} f_j = \sum f$. In this form it is useful for grouped data.

Sometimes the standard deviation of a sample's data is defined with $(N - 1)$ replacing N in the denominators of the expressions in equations (5) and (6) because the resulting value represents a better estimate of the standard deviation of a population from which the sample is taken. For large values of N (certainly $N > 30$), there is practically no difference between the two definitions. Also, when the better estimate is needed we can always obtain it by multiplying the standard deviation computed according to the first definition by $\sqrt{N/(N-1)}$. Hence we shall adhere to the form of equations (5) and (6).

The Variance

The *variance* of a set of data is defined as the square of the standard deviation and is thus given by s^2 in equations (5) and (6).

When it is necessary to distinguish the standard deviation of a population from the standard deviation of a sample drawn from this population, we often use the symbol s for the latter and σ (lowercase Greek *sigma*) for the former. Thus s^2 and σ^2 would represent the *sample variance* and *population variance*, respectively.

Properties of the Standard Deviation

1. The standard deviation can be defined as

$$ s = \sqrt{\frac{\sum\limits_{j=1}^{N} (X_j - a)^2}{N}} $$

where a is an average besides the arithmetic mean. Of all such standard deviations, the minimum is that for which $a = X$.

2. For normal distributions (see Chapter 4), it turns out that (as shown in Fig. 2-1):

   (a) 68.27 % of the cases are included between $X - s$ and $X + s$ (i.e., one standard deviation on either side of the mean).

   (b) 95.45% of the cases are included between $X - 2s$ and $X + 2s$ (i.e., two standard deviations on either side of the mean).

   (c) 99.73% of the cases are included between $X - 3s$ and $X + 3s$ (i.e., three standard deviations on either side of the mean).

   For moderately skewed distributions, the above percentages may hold approximately (unlike the normal distribution which is symmetrical, skewed distributions are asymmetrical).

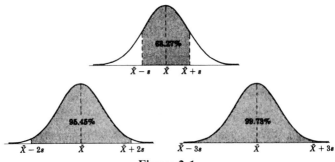

Figure 2-1

Absolute and Relative Dispersion; Coefficient of Variation

The actual variation, or dispersion, as determined from the standard deviation or other measure of dispersion is called the *absolute dispersion*. However, a variation (or dispersion) of 10 inches (in) in measuring a distance of 1,000 feet (ft) is quite different in effect from the same variation of 10 in in a distance of 20 ft. A measure of this effect is supplied by the *relative dispersion*, which is defined by

$$\text{Relative dispersion} = \text{absolute dispersion} \,/\, \text{average} \qquad (7)$$

If the absolute dispersion is the standard deviation s and if the average is the mean , then the relative dispersion is called the *coefficient of variation*, or *coefficient of dispersion*; it is denoted by V and is given by

$$\text{Coefficient of variation } (V) = s\,/\,X \qquad (8)$$

and is generally expressed as a percentage.

Note that the coefficient of variation is independent of the units used. For this reason, it is useful in comparing distributions where the units may be different. A disadvantage of the coefficient of variation is that it fails to be useful when X is close to zero.

Standardized Variable: Standard Scores

The variable that measures the deviation from the mean in units of the standard deviation is called a *standardized variable*, is a dimension less quantity (i.e., is independent of the units used), and is given by

$$Z = (X_1 - X)\,/\,s \qquad (9)$$

If the deviations from the mean are given in units of the standard deviation, they are said to be expressed in *standard units*, or *standard scores*. These are of great value in the comparison of distributions.

Chapter 3
ELEMENTARY
PROBABILITY
THEORY

Definitions of Probability

Classic Definition

Suppose that an event E can happen in h ways out of a total of n possible equally likely ways. Then the probability of occurrence of the event (called its *success*) is denoted by

$$p = \Pr\{E\} = h / n$$

The probability of nonoccurrence of the event (called its *failure*) is denoted by

$$q = \Pr\{\text{not } E\} = (n - h) / n = 1 - (h / n) = 1 - p = 1 - \Pr\{E\}$$

Thus $p + q = 1$, or $\Pr\{E\} + \Pr\{\text{not } E\} = 1$. The event "not E" is sometimes denoted by $\sim E$.

EXAMPLE 3.1. Let E be the event that the number 3 or 4 turns up in a single toss of a die. There are six ways in which the die can fall, resulting in the numbers 1, 2, 3, 4, 5, or 6; and if the die is *fair* (i.e., not *loaded*), we can *assume* these six ways to be equally likely. Since E can occur in two of these ways, we have $p = \Pr\{E\} = 2 / 6 = 1 / 3$.

The probability of not getting a 3 or 4 (i.e., getting a 1, 2, 5, or 6) is $q = \Pr\{\sim E\} = 1 - 1 / 3 = 2 / 3$.

Note that the probability of an event is a number between 0 and 1. If the event cannot occur, its probability is 0. If it must occur (i.e., its occurrence is *certain*), its probability is 1.

If p is the probability that an event will occur, the *odds* in favor of its happening are $p : q$ (read "p to q"); the odds against its happening are $q : p$. Thus the odds against a 3 or 4 in a single toss of a fair die are $q : p = : = 2 : 1$ (i.e., 2 to 1).

Relative-Frequency Definition

The classic definition of probability has a disadvantage in that the words "equally likely" are vague. In fact, since these words seem to be

synonymous with "equally probable," the definition is *circular* because we are essentially defining probability in terms of itself. For this reason,
a statistical definition of probability has been
advocated by some people. According to this the
estimated probability, or *empirical probability*,
of an event is taken to be the *relative frequency* of
occurrence of the event when the number of
observations is very large. The probability itself
is the *limit* of the relative frequency as the number of observations increases indefinitely.

EXAMPLE 3.2. If 1000 tosses of a coin result in 529 heads, the relative frequency of heads is 529 / 1000 = 0.529. If another 1000 tosses
results in 493 heads, the relative frequency in the total of 2000 tosses is
(529 + 493) / 2000 = 0.511. According to the statistical definition, by
continuing in this manner we should ultimately get closer and closer to
a number that represents the probability of a head in a single toss of the
coin. From the results so far presented, this should be 0.5.

The statistical definition, although useful in practice, has difficulties
from a mathematical point of view, since an actual limiting number may
not really exist. For this reason, modern probability theory has been developed *axiomatically*; that is, the theory leaves the concept of probability
undefined, much the same as *point* and *line* are undefined in geometry.

Conditional Probability: Independent and Dependent Events

If E_1 and E_2 are two events, the probability that E_2 occurs given that E_1
has occurred is denoted by $\Pr\{E_2 \mid E_1\}$ or $\Pr\{E_2$ given $E_1\}$, and is called
the *conditional probability* of E_2 given that E_1, has occurred.

If the occurrence or nonoccurrence of E_1 does not affect the probability of occurrence of E_2, then $\Pr\{E_2 \mid E_1\} = \Pr\{E_2\}$ and we say that E_1
and E_2 are *independent events*; otherwise, they are *dependent events*.

If we denote by E_1E_2 the event that "both E_1 and E_2 occur," sometimes called a *compound event*, then

$$\Pr\{E_1 E_2\} = \Pr\{E_1\}\Pr\{E_2 \mid E_1\} \qquad (1)$$

In particular,

$$\Pr\{E_1 E_2\} = \Pr\{E_1\}\,\Pr\{E_2\} \text{ for independent events} \qquad (2)$$

For three events E_1, E_2, and E_3, we have

$$\Pr\{E_1 E_2 E_3\} = \Pr\{E_1\}\Pr\{E_2 \mid E_1\}\Pr\{E_3 \mid E_1 E_2\} \qquad (3)$$

That is, the probability of occurrence of E_1, E_2, E_3 is equal to (the probability of E_1) × (the probability of E_2 given that E_1 has occurred) × (the probability of E_3 given that both E_1 and E_2 have occurred). In particular,

$$\Pr\{E_1 E_2 E_3\} = \Pr\{E_1\}\Pr\{E_2\}\Pr\{E_3\} \text{ for independent events} \qquad (4)$$

In general, if E_1, E_2, E_3,..., E_n are n independent events having respective probabilities p_1, p_2, p_3,..., p_n, then the probability of occurrence of E_1 and E_2 and E_3 and ... E_n is $p_1 p_2 p_3 \cdots p_n$.

EXAMPLE 3.3. Let E_1 and E_2 be the events "heads on fifth toss" and "heads on sixth toss" of a coin, respectively. Then E_1 and E_2 are independent events, and thus the probability of heads on both the fifth and sixth tosses is (assuming the coin to be fair)

$$\Pr\{E_1 E_2\} = \Pr\{E_1\}\Pr\{E_2\} = (1\,/\,2)(1\,/\,2) = 1\,/\,4$$

EXAMPLE 3.4. If the probability that A will be alive in 20 years is 0.7 and the probability that B will be alive in 20 years is 0.5, then the probability that they will both be alive in 20 years is $(0.7)(0.5) = 0.35$.

EXAMPLE 3.5. Suppose that a box contains 3 white balls and 2 black balls. Let E_1 be the event "first ball drawn is black" and E_2 the event "second ball drawn is black," where the balls are not replaced after being drawn. Here E_1 and E_2 are dependent events.

The probability that the first ball drawn is black is $\Pr\{E_1\} = 2\,/\,(3+2) = 2\,/\,5$. The probability that the second ball drawn is black, given that the first ball drawn was black, is $\Pr\{E_2 \mid E_1\} = 1\,/\,(3+1) = 1\,/\,4$. Thus the probability that both balls drawn are black is

$$\Pr\{E_1\ E_2\} = \Pr\{E_1\}\Pr\{E_2|E_1\} = (2\ /\ 5) \bullet (1\ /\ 4) = 1\ /\ 10$$

Mutually Exclusive Events

Two or more events are called *mutually exclusive* if the occurrence of any one of them excludes the occurrence of the others. Thus if E_1 and E_2 are mutually exclusive events, then $\Pr\{E_1\ E_2\} = 0$.

If $E_1 + E_2$ denotes the event that "either E_1 or E_2 or both occur," then

$$\Pr\{E_1 + E_2\} = \Pr\{E_2\} + \Pr\{E_1\ E_2\} - \Pr\{E_1\ E_2\} \qquad (5)$$

In particular,

$$\Pr\{E_1 + E_2\} = \Pr\{E_1\} + \Pr\{E_2\} \quad \text{for mutually exclusive events} \quad (6)$$

As an extension of this, if E_1, E_2,..., E_n are n mutually exclusive events having respective probabilities of occurrence $p_1, p_2,..., p_n$ then the probability of occurrence of either E_1 or E_2 or ... E_n is $p_1 + p_2 + ... + p_n$.

Equation (5) can also be generalized to three or more mutually exclusive events.

EXAMPLE 3.6. If E_1 is the event "drawing an ace from a deck of cards" and E_2 is the event "drawing a king," then $\Pr\{E_1\} = 4\ /\ 52 = 1\ /\ 13$ and $\Pr\{E_2\} = 4\ /\ 52 = 1\ /\ 13$. The probability of drawing either an ace or a king in a single draw is

$$\Pr\{E_1 + E_2\} = \Pr\{E_1\} + \Pr\{E_2\} = 1\ /\ 13 + 1\ /\ 13 = 2\ /\ 13$$

since both an ace and a king cannot be drawn in a single draw and are thus mutually exclusive events.

EXAMPLE 3.7. If E_1 is the event "drawing an ace" from a deck of cards and E_2 is the event "drawing a spade," then E_1 and E_2 are not mutually exclusive since the ace of spades can be drawn. Thus the probability of drawing either an ace or a spade or both is

$$\begin{aligned}\Pr\{E_1 + E_2\} &= \Pr\{E_1\} + \Pr\{E_2\} - \Pr\{E_1\ E_2\} \\ &= 4\ /\ 52 + 13\ /\ 52 - 1\ /\ 52 = 16\ /\ 52 = 4\ /\ 13\end{aligned}$$

Probability Distributions

Discrete

If a variable X can assume a discrete set of values X_1, $X_2,..., X_K$ with respective probabilities p_1, $p_2,..., p_K$ where $p_1 + p_2 + ... + p_K = 1$, we say that a *discrete probability distribution* for X has been defined. The function $p(X)$, which has the respective values $p_1, p_2,..., p_K$ for $X = X_1, X_2,..., X_K$, is called the *probability function*, or *frequency function*, of X. Because X can assume certain values with given probabilities, it is often called a *discrete random variable*. A random variable is also known as a *chance variable* or *stochastic variable*.

EXAMPLE 3.8. Let a pair of fair dice be tossed and let X denote the sum of the points obtained. Then the probability distribution is as shown in Table 3-1. For example, the probability of getting sum 5 is 4 / 36 = 1 / 9; thus in 900 tosses of the dice we would expect 100 tosses to give the sum 5.

Table 3-1

X	2	3	4	5	6	7	8	9	10	11	12
p(X)	1/36	2/36	3/36	4/36	5/36	6/36	5/36	4/36	3/36	2/36	1/36

Note that this is analogous to a relative-frequency distribution with probabilities replacing the relative frequencies. Thus we can think of probability distributions as theoretical or ideal limiting forms of relative-frequency distributions when the number of observations made is very large. For this reason, we can think of probability distributions as being distributions of *populations*, whereas relative-frequency distributions are distributions of *samples* drawn from this population.

The probability distribution can be represented graphically by plotting $p(X)$ against X, just as for relative-frequency distributions.

You Need to Know

By cumulating probabilities, we obtain *cumulative probability distributions*, which are analogous to cumulative relative-frequency distributions. The function associated with this distribution is sometimes called a *distribution function*.

Continuous

The above ideas can be extended to the case where the variable X may assume a continuous set of values. The relative-frequency polygon of a sample becomes, in the theoretical or limiting case of a population, a continuous curve (such as shown in Fig. 3-1) whose equation is $Y = p(X)$. The total area under this curve bounded by the X axis is equal to 1, and the area under the curve between lines $X = a$ and $X = b$ (shaded in Fig. 3-1) gives the probability that X lies between a and b, which can be denoted by $\Pr\{a < X < b\}$.

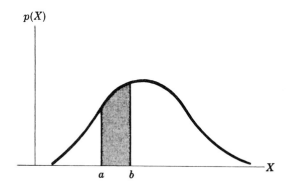

Figure 3-1

We call $p(X)$ a *probability density function*, or briefly a *density function*, and when such a function is given, we say that a *continuous probability distribution* for X has been defined. The variable X is then often called a *continuous random variable*.

As in the discrete case, we can define cumulative probability distributions and the associated distribution functions.

Mathematical Expectation

If p is the probability that a person will receive a sum of money S, the *mathematical expectation* (or simply the *expectation*) is defined as pS.

EXAMPLE 3.9. If the probability that a man wins a \$10 prize is 1 / 5, his expectation is (\$10) = \$2.

The concept of expectation is easily extended. If X denotes a discrete random variable that can assume the values X_1, X_2,..., X_K with respective probabilities p_1, p_2,..., p_K, where $p_1 + p_2 + \ldots + p_K = 1$, the *mathematical expectation* of X (or simply the *expectation* of X), denoted by $E(X)$, is defined as

$$E(X) = p_1X_1 + p_2X_2 + p_KX_K = \Sigma_{j=1}^{K} p_jX_j = \Sigma \ pX \qquad (7)$$

If the probabilities p_j in this expectation are replaced with the relative frequencies f_j / N, where $N = \Sigma \ f_j$, the expectation reduces to $(\Sigma \ fX) / N$, which is the arithmetic mean of a sample of size N in which X_1, X_2,..., X_K appear with these relative frequencies. As N gets larger and larger, the relative frequencies f_j / N approach the probabilities p_j. Thus we are led to the interpretation that $E(X)$ represents the mean of the population from which the sample is drawn. We denote the population mean by the Greek letter μ (mu).

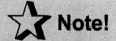 **Note!**

Expectation can also be defined for continuous random variables, but the definition requires the use of calculus.

Relation between Population, Sample Mean, and Variance

If we select a sample of size N at random from a population (i.e., we assume that all such samples are equally probable), then it is possible to show that the *expected value of the sample mean m is the population mean μ.*

It does not follow, however, that the expected value of any quantity computed from a sample is the corresponding population quantity. For example, the expected value of the sample variance as we have defined it is not the population variance, but $(N - 1)/N$ times this variance. This is why some statisticians choose to define the sample variance as our variance multiplied by $N/(N - 1)$.

Combinatorial Analysis

In obtaining probabilities of complex events, an enumeration of cases is often difficult, tedious, or both. To facilitate the labor involved, use is made of basic principles studied in a subject called *combinatorial analysis.*

Fundamental Principle

If an event can happen in any one of n_1 ways, and if when this has occurred another event can happen in any one of n_2 ways, then the number of ways in which both events can happen in the specified order is $n_1 n_2$.

Factorial *n*

Factorial n, denoted by $n!$, is defined as

$$n! = n(n - 1)(n - 2) \ldots 1 \tag{8}$$

Thus $5! = 5 \cdot 4 \cdot 3 \cdot 2 \cdot 1 = 120$, and $4!3! = (4 \cdot 3 \cdot 2 \cdot 1)(3 \cdot 2 \cdot 1) = 144$. It is convenient to define $0! = 1$.

Permutations

A permutation of n different objects taken r at a time is an *arrangement* of r out of the n objects, with attention given to the order of arrangement. The number of permutations of n objects taken r at a time is denoted by $_nP_r$, $P(n, r)$, or $P_{n,r}$ and is given by

$$_nP_r = n(n - 1)(n - 2) \ldots (n - r + 1) = n! / (n - r)! \qquad (9)$$

In particular, the number of permutations of n objects taken n at a time is

$$_nP_r = n (n - 1)(n - 2) \ldots 1 = n!$$

EXAMPLE 3.10. The number of permutations of the letters a, b, and c taken two at a time is $_3P_2 = 3 \cdot 2 = 6$. These are ab, ba, ac, ca, bc, and cb.

The number of permutations of n objects consisting of groups of which n_1 are alike, n_2 are alike, ... is

$$n! / (n_1! \, n_2! \, \ldots) \quad \text{where } n = n_1 + n_2 + \ldots n_n \qquad (10)$$

EXAMPLE 3.11. The number of permutations of letters in the word *statistics* is

$$10! / (3!3!1!2!1!) = 50,400$$

since there are $3s$'s, $3t$'s, $1a$, $2i$'s, and $1c$.

Combinations

A combination of n different objects taken r at a time is a selection of r out of the n objects, with no attention given to the order of arrangement. The number of combinations of n objects taken r at a time is denoted by the symbol $\binom{n}{r}$ and is given by

$$\binom{n}{r} = \frac{n(n-1) \cdots (n-r+1)}{r!} = \frac{n!}{r!(n-r)!} \qquad (11)$$

EXAMPLE 3.12. The number of combinations of the letters a, b, and c taken two at a time is

$$\binom{3}{2} = \frac{3 \cdot 2}{2!} = 3$$

These are ab, ac, and bc. Note that ab is the same combination as ba, but not the same permutation.

Relation of Probability to Point Set Theory

In modern probability theory, we think of all possible outcomes (or results) of an experiment, game, etc., as points in a space (which can be of one, two, three, etc., dimensions), called a *sample space S*. If S contains only a finite number of points, then with each point we can associate a nonnegative number, called a *probability*, such that the sum of all numbers corresponding to all points in S add to 1. An event is a *set* (or *collection*) of points in S, such as indicated by E_1 or E_2 in Fig. 3-2; this figure is called an *Euler diagram* or *Venn diagram*.

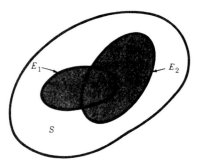

Figure 3-2

The event $E_1 + E_2$ is the set of points that are *either in E_1 or E_2 or both*, while the event $E_1 E_2$ is the set of points *common to both E_1 and E_2*. Thus the probability of an event such as E_1 is the sum of the probabilities associated with all points contained in the set E_1. Similarly, the probability of $E_1 + E_2$, denoted by $\Pr\{E_1 + E_2\}$, is the sum of the probabilities associated with all points contained in the set $E_1 + E_2$. If E_1 and E_2 have no points in common (i.e., the events are mutually exclusive), then $Pr\{E_1 + E_2\} = Pr\{E_1\} + Pr\{E_2\}$. If they have points in common, then $Pr\{E_1 + E_2\} = Pr\{E_1\} + Pr\{E_2\} - Pr\{E_1 E_2\}$.

The set $E_1 + E_2$ is sometimes denoted by $E_1 \cup E_2$ and is called the *union* of the two sets. The set $E_1 E_2$ is sometimes denoted by $E_1 \cap E_2$ and is called the *intersection* of the two sets. Extensions to more than two sets can be made; thus instead of $E_1 + E_2 + E_3$ and $E_1 E_2 E_3$, we could use the notations $E_1 \cup E_2 \cup E_3$ and $E_1 \cap E_2 \cap E_3$, respectively.

The symbol φ (the Greek letter *phi*) is sometimes used to denote a set with no points in it, called the *null set*. The probability associated with an event corresponding to this set is zero (i.e., $Pr\{\varphi\} = 0$). If E_1 and E_2 have no points in common, we can write $E_1 E_2 = \varphi$, which means that the corresponding events are mutually exclusive, whereby $Pr\{E_1 E_2\} = 0$.

With this modern approach, a random variable is a function defined at each point of the sample space. In the case where S has an infinite number of points, the above ideas can be extended by using concepts of calculus.

Chapter 4
THE BINOMIAL, NORMAL, AND POISSON DISTRIBUTIONS

IN THIS CHAPTER:

✔ *The Binomial Distribution*
✔ *The Normal Distribution*
✔ *The Poisson Distribution*
✔ *The Multinomial Distribution*

The Binomial Distribution

If p is the probability that an event will happen in any single trial (called the probability of a *success*) and $q = 1 - p$ is the probability that it will fail to happen in any single trial (called the probability of a *failure*), then the probability that the event will happen exactly X times in N trials (i.e., X successes and $N - X$ failures will occur) is given by

$$p(X) = \binom{N}{X} p^X q^{N-X} = \frac{N!}{X!(N-X)!} p^X q^{N-X} \qquad (1)$$

where $X = 0, 1, 2,..., N$; $N! = N(N - 1)(N - 2) \ldots 1$; and $0! = 1$ by definition.

EXAMPLE 4.1. The probability of getting exactly 2 heads in 6 tosses of a fair coin is

$$\binom{6}{2}\left(\frac{1}{2}\right)^2\left(\frac{1}{2}\right)^{6-2} = \frac{6}{2!\,4!}\left(\frac{1}{2}\right)^6 = \frac{15}{64}$$

using formula (1) with $N = 6$, $X = 2$, and $p = q = 1/2$.

EXAMPLE 4.2. The probability of getting at least 4 heads in 6 tosses of a fair coin is

$$\binom{6}{4}\left(\frac{1}{2}\right)^4\left(\frac{1}{2}\right)^{6-4} + \binom{6}{5}\left(\frac{1}{2}\right)^5\left(\frac{1}{2}\right)^{6-5} + \binom{6}{6}\left(\frac{1}{2}\right)^6\left(\frac{1}{2}\right)^{6-6} = \frac{15}{64} + \frac{6}{64} + \frac{1}{64} = \frac{11}{32}$$

The discrete probability distribution (1) is often called the *binomial distribution* since for $X = 0, 1, 2, \ldots, N$ it corresponds to successive terms of the *binomial formula*, or *binomial expansion*,

$$(q + p)^N = q^N + \binom{N}{1}q^{N-1}p + \binom{N}{2}q^{N-2}p^2 + \cdots + p^N \tag{2}$$

where 1, $\binom{N}{1}$, $\binom{N}{2}$,... are called the *binomial coefficients*.

EXAMPLE 4.3.

$$(q + p)^4 = q^4 + \binom{4}{1}q^3p + \binom{4}{2}q^2p^2 + \binom{4}{3}qp^3 + p^4$$

Distribution (1) is also called the *Bernoulli distribution* after James Bernoulli, who discovered it at the end of the seventeenth century. Some properties of the binomial distribution are listed in Table 4-1.

Table 4-1 Binomial Distribution

Mean	$\mu = Np$
Variance	$\sigma^2 = Npq$
Standard deviation	$\sigma = \sqrt{Npq}$

EXAMPLE 4.4. In 100 tosses of a fair coin the mean number of heads is $\mu = Np = (100)(\frac{1}{2}) = 50$; this is the *expected* number of heads in 100 tosses of the coin. The standard deviation is $\sigma = \sqrt{Npq} = \sqrt{(100)\,(\frac{1}{2})(\frac{1}{2})} = 5$.

The Normal Distribution

One of the most important examples of a continuous probability distribution is the *normal distribution, normal curve*, or *gaussian distribution*. It is defined by the equation

$$Y = \frac{1}{\sigma\sqrt{2\pi}}\, e^{-\frac{1}{2}(X-\mu)^2/\sigma^2} \qquad (3)$$

where μ = mean, σ = standard deviation, $\pi = 3.14159...$, and $e = 2.71828....$ The total area bounded by curve (3) and the X axis is 1; hence the area under the curve between two ordinates $X = a$ and $X = b$, where $a < b$, represents the probability that X lies between a and b. This probability is denoted by $\Pr\{a < X < b\}$.

When the variable X is expressed in terms of standard units [$z = (X - \mu)\,/\,\sigma$], equation (3) is replaced by the so-called *standard form*

$$Y = \frac{1}{\sqrt{2\pi}}\, e^{-\frac{1}{2}z^2} \qquad (4)$$

In such case we say that z is *normally distributed with mean 0 and variance 1*. Figure 4-1 is a graph of this standardized normal curve. It shows that the areas included between $z = -1$ and $+1$, $z = -2$ and $+2$, and $z = -3$ and $+3$ are equal, respectively, to 68.27%, 95.45%, and 99.73% of the total area, which is 1. The table in Appendix A shows the areas under this curve bounded by the ordinates at $z = 0$ and any positive value of z. From this table the area between any two ordinates can be found by using the symmetry of the curve about $z = 0$.

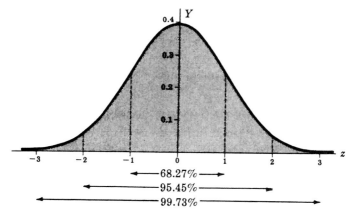

Figure 4-1

Some properties of the normal distribution given by equation (*3*) are listed in Table 4-2.

Table 4-2 Normal Distribution

Mean	μ
Variance	σ^2
Standard deviation	σ

Relation between the Binomial and Normal Distributions

If N is large and if neither p nor q is too close to zero, the binomial distribution can be closely approximated by a normal distribution with standardized variable given by

$$z = \frac{X - Np}{\sqrt{Npq}}$$

The approximation becomes better with increasing N, and in the limiting case it is exact. In practice the approximation is very good if both Np and Nq are greater than 5.

The Poisson Distribution

The discrete probability distribution

$$p(X) = \frac{\lambda^X e^{-\lambda}}{X!} \qquad X = 0, 1, 2, \ldots \tag{5}$$

where $e = 2.71828 \ldots$ and λ is a given constant, is called the *Poisson distribution* after Siméon-Denis Poisson, who discovered it in the early part of the nineteenth century.

Some properties of the Poisson distribution are listed in Table 4-3.

Table 4-3 Poisson's Distribution

Mean	$\mu = \lambda$
Variance	$\sigma^2 = \lambda$
Standard deviation	$\sigma = \sqrt{\lambda}$

Relation between the Binomial and Poisson Distributions

In the binomial distribution (*1*), if N is large while the probability p of the occurrence of an event is close to 0, so that $q = 1 - p$ is close to 1, the event is called a *rare event*. In practice we shall consider an event to be rare if the number of trials is at least 50 ($N \geq 50$) while Np is less than 5. In such case the binomial distribution (*1*) is very closely approximated by the Poisson distribution (*5*) with $\sigma\lambda = Np$. This is indicated by comparing Tables 4-1 and 4-3 — for by placing $\lambda = Np$, $q \approx 1$, and $p \approx 0$ in Table 4-1, we get the results in Table 4-3.

Since there is a relation between the binomial and normal distributions, it follows that there also is a relation between the Poisson and normal distributions.

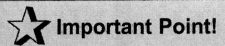

Important Point!

It can in fact be shown that the Poisson distribution approaches a normal distribution with standardized variable $(X - A)$ $\sqrt{\lambda}$ as λ increases indefinitely.

The Multinomial Distribution

If events $E_1, E_2,..., E_K$ can occur with probabilities $p_1, p_2,..., p_K$, respectively, then the probability that $E_1, E_2,..., E_K$ will occur $X_1, X_2,..., X_K$ times, respectively, is

$$\frac{N!}{X_1! X_2! \cdots X_K!} p_1^{X_1} p_2^{X_2} \cdots p_K^{X_K} \tag{6}$$

where $X_1 + X_2 + ... + X_K = N$. This distribution, which is a generalization of the binomial distribution, is called the *multinomial distribution* since equation (6) is the general term in the *multinomial expansion* $(p_1 + p_2 + ... + p_K)^N$.

EXAMPLE 4.5. If a fair die is tossed 12 times, the probability of getting 1, 2, 3, 4, 5, and 6 points exactly twice each is

$$\frac{12!}{2!2!2!2!2!2!} \left(\frac{1}{6}\right)^2 \left(\frac{1}{6}\right)^2 \left(\frac{1}{6}\right)^2 \left(\frac{1}{6}\right)^2 \left(\frac{1}{6}\right)^2 \left(\frac{1}{6}\right)^2 = \frac{1925}{559,872} = 0.00344$$

The *expected* numbers of times that $E_1, E_2,..., E_K$ will occur in N trials are $Np_1, Np_2,..., Np_K$, respectively.

Chapter 5
ELEMENTARY
SAMPLING THEORY

IN THIS CHAPTER:

✔ *Sampling Theory*
✔ *Random Samples and*
   Random Numbers
✔ *Sampling with and*
   without Replacement
✔ *Sampling Distributions*
✔ *Sampling Distribution of Means*
✔ *Sampling Distribution*
   of Proportions
✔ *Sampling Distributions of*
   Differences and Sums
✔ *Standard Errors*

Sampling Theory

Sampling theory is a study of relationships existing between a population and samples drawn from the population. It is of great value in many connections. For example, it is useful in *estimating* unknown population quantities (such as population mean and variance), often called *population parameters* or briefly *parameters*, from a knowledge of corresponding sample quantities (such as sample mean and variance), often called *sample statistics* or briefly *statistics*.

Sampling theory is also useful in determining whether the observed differences between two samples are due to chance variation or whether they are really significant. Such questions arise, for example, in testing a new serum for use in the treatment of a disease or in deciding whether one production process is better than another. Their answers involve the use of so-called *tests of significance and hypotheses* that are important in the *theory of decisions.*

In general, a study of the inferences made concerning a population by using samples drawn from it, together with indications of the accuracy of such inferences by using probability theory, is called *statistical inference.*

Random Samples and Random Numbers

In order that the conclusions of sampling theory and statistical inference be valid, samples must be chosen so as to be *representative* of a population. A study of sampling methods and of the related problems that arise is called the *design of the experiment.*

One way in which a representative sample may be obtained is by a process called *random sampling*, according to which each member of a population has an equal chance of being included in the sample. One technique for obtaining a random sample is to assign numbers to each member of the population, write these numbers on small pieces of paper, place them in an urn, and then draw numbers from the urn, being careful to mix thoroughly before each drawing. An alternative method is to use a table of *random numbers* specially constructed for such purposes.

Sampling with and without Replacement

If we draw a number from an urn, we have the choice of replacing or not replacing the number into the urn before a second drawing. In the first case the number can come up again and again, whereas in the second it can only come up once. Sampling where each member of the population may be chosen more than once is called *sampling with replacement*, while if each member cannot be chosen more than once, it is called *sampling without replacement*.

Populations are either finite or infinite. If, for example, we draw 10 balls successively without replacement from an urn containing 100 balls, we are sampling from a finite population; while if we toss a coin 50 times and count the number of heads, we are sampling from an infinite population.

A finite population in which sampling is with replacement can theoretically be considered infinite, since any number of samples can be drawn without exhausting the population.

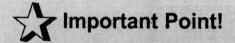 **Important Point!**

For many practical purposes, sampling from a finite population that is very large can be considered to be sampling from an infinite population.

Sampling Distributions

Consider all possible samples of size N that can be drawn from a given population (either with or without replacement). For each sample, we can compute a statistic (such as the mean and the standard deviation) that will vary from sample to sample. In this manner we obtain a distribution of the statistic that is called its *sampling distribution.*

If, for example, the particular statistic used is the sample mean, then the distribution is called the *sampling distribution of means*, or the *sampling distribution of the mean.* Similarly, we could have sampling distributions of standard deviations, variances, medians, proportions, etc.

For each sampling distribution, we can compute the mean, standard deviation, etc. Thus we can speak of the mean and standard deviation of the sampling distribution of means, etc.

Sampling Distribution of Means

Suppose that all possible samples of size N are drawn without replacement from a finite population of size $N_p > N$. If we denote the mean and standard deviation of the sampling distribution of means by $\mu_{\bar{x}}$ and $\sigma_{\bar{x}}$ and the population mean and standard deviation by μ and σ, respectively, then

$$\mu_{\bar{x}} = \mu \quad \text{and} \quad \sigma_{\bar{x}} = \frac{\sigma}{\sqrt{N}} \sqrt{\frac{N_p - N}{N_p - 1}} \tag{1}$$

If the population is infinite or if sampling is with replacement, the above results reduce to

$$\mu_{\bar{x}} = \mu \quad \text{and} \quad \sigma_{\bar{x}} = \frac{\sigma}{\sqrt{N}} \tag{2}$$

For large values of N ($N \geq 30$), the sampling distribution of means is approximately a normal distribution with mean $\mu_{\bar{x}}$ and standard deviation $\sigma_{\bar{x}}$, irrespective of the population (so long as the population mean and variance are finite and the population size is at least twice the sample size). This result for an infinite population is a special case of the *central limit theorem* of advanced probability theory, which shows that the accuracy of the approximation improves as N gets larger. This is sometimes indicated by saying that the sampling distribution is *asymptotically normal.*

In case the population is normally distributed, the sampling distribution of means is also normally distributed even for small values of N (i.e., $N < 30$).

Sampling Distribution of Proportions

Suppose that a population is infinite and that the probability of occurrence of an event (called its success) is p, while the probability of nonoccurrence of the event is $q = 1 - p$. For example, the population may be all possible tosses of a fair coin in which the probability of the event "heads" is $p = 1 / 2$. Consider all possible samples of size N drawn from this population, and for each sample determine the proportion P of successes. In the case of the coin, P would be the proportion of heads turning up in N tosses. We thus obtain a *sampling distribution of proportions* whose mean μ_P and standard deviation σ_P are given by

$$\mu_P = p \quad \text{and} \quad \sigma_P = \sqrt{\frac{pq}{N}} = \sqrt{\frac{p(1-p)}{N}} \tag{3}$$

which can be obtained from equations (2) by placing $\mu = p$ and $\sigma = \sqrt{pq}$. For large values of N ($N \geq 30$), the sampling distribution is very closely normally distributed. Note that the population is *binomially distributed*.

Equations (3) are also valid for a finite population in which sampling is with replacement. For finite populations in which sampling is without replacement, equations (3) are replaced by equations (1) with $\mu = p$ and $\sigma = \sqrt{pq}$.

Note that equations (3) are obtained most easily by dividing the mean and standard deviation (Np and $\sqrt{pq}$) of the binomial distribution by N.

Sampling Distributions of Differences and Sums

Suppose that we are given two populations. For each sample of size N_1 drawn from the first population, let us compute a statistic S_1; this yields a sampling distribution for the statistic S_1, whose mean and standard deviation we denote by μ_{S1}, and σ_{S1}, respectively. Similarly, for each sample of size N_2 drawn from the second population, let us compute a statistic S_2; this yields a sampling distribution for the statistic S_2, whose mean and standard deviation are denoted by μ_{S2} and σ_{S2}. From all possible combinations of these samples from the two populations we can obtain a distribution of the differences, $S_1 - S_2$, which is called the *sampling distribution of differences of the statistics*. The mean and standard deviation of this sampling distribution, denoted respectively by μ_{S1-S2} and σ_{S1-S2}, are given by

$$\mu_{S1-S2} = \mu_{S1} - \mu_{S2} \quad \text{and} \quad \sigma_{S1-S2} = \sqrt{\sigma_{S1}^2 + \sigma_{S2}^2} \tag{4}$$

provided that the samples chosen do not in any way depend on each other (i.e., the samples are *independent*).

If S_1 and S_2 are the sample means from the two populations — which means we denote by μ_1 and μ_2 respectively — then the sampling distribution of the differences of means is given for infinite populations with means and standard deviations (μ_1, σ_1) and (μ_2, σ_2), respectively, by

$$\mu_{\bar{X}1-\bar{X}2} = \mu_{\bar{X}1} - \mu_{\bar{X}2} = \mu_1 - \mu_2$$

and

$$\sigma_{\bar{X}1-\bar{X}2} = \sqrt{\sigma_{\bar{X}1}^2 + \sigma_{\bar{X}2}^2} = \sqrt{\frac{\sigma_1^2}{N_1} + \frac{\sigma_2^2}{N_2}} \tag{5}$$

using equations (2). The result also holds for finite populations if sampling is with replacement. Similar results can be obtained for finite populations in which sampling is without replacement by using equations (1).

Corresponding results can be obtained for the sampling distributions of differences of proportions from two binomially distributed populations with parameters (p_1, q_1) and (p_2, q_2), respectively. In this case S_1 and S_2 correspond to the proportion of successes, P_1 and P_2, and equations (4) yield the results

$$\mu_{P1-P2} = \mu_{P1} - \mu_{P2} = p_1 - p_2$$

and

$$\sigma_{P1-P2} = \sqrt{\sigma_{P1}^2 + \sigma_{P2}^2} = \sqrt{\frac{p_1 q_1}{N_1} + \frac{p_2 q_2}{N_2}} \tag{6}$$

If N_1 and N_2 are large (N_1, $N_2 \geq 30$), the sampling distributions of differences of means or proportions are very closely normally distributed.

It is sometimes useful to speak of the *sampling distribution of the sum of statistics*. The mean and standard deviation of this distribution are given by

$$\mu_{S1+S2} = \mu_{S1} + \mu_{S2} \quad \text{and} \quad \sigma_{S1+S2} = \sqrt{\sigma_{S1}^2 + \sigma_{S2}^2} \tag{7}$$

assuming that the samples are *independent*.

Standard Errors

The standard deviation of a sampling distribution of a statistic is often called its *standard error*. Table 5-1 lists standard errors of sampling distributions for various statistics under the conditions of random sampling from an infinite (or very large) population or of sampling with replacement from a finite population.

The quantities μ, σ, p and , s, P denote, respectively, the population and sample means, standard deviations, and proportions.

Table 5-1 Standard Errors for Some Sampling Distributions

Sampling Distribution	Standard Error
Means	$\sigma_{\bar{x}} = \dfrac{\sigma}{\sqrt{N}}$
Proportions	$\sigma_P = \sqrt{\dfrac{p(1-p)}{N}} = \sqrt{\dfrac{pq}{N}}$
Standard deviation	$(1) \ \sigma_s = \dfrac{\sigma}{\sqrt{2N}}$
	$(2) \ \sigma_s = \sqrt{\dfrac{\mu_4 - \mu_2^2}{4N\mu_2}}$
Variances	$\sigma_{med} = \sigma\sqrt{\dfrac{\pi}{2N}} = \dfrac{1.2533\sigma}{\sqrt{N}}$

It is noted that if the sample size N is large enough, the sampling distributions are normal or nearly normal. For this reason, the methods are known as *large sampling methods*. When $N < 30$, samples are called *small*.

When population parameters such as μ or p are unknown, they may be estimated closely by their corresponding sample statistics namely, s (or $\hat{s} = \sqrt{N / (N-1)} \ s$) and P — if the samples are large enough.

Chapter 6
STATISTICAL ESTIMATION THEORY

IN THIS CHAPTER:

45

Estimation of Parameters

In the last chapter we saw how sampling theory can be employed to obtain information about samples drawn at random from a known population. From a practical viewpoint, however, it is often more important to be able to infer information about a population from samples drawn from it. Such problems are dealt with in *statistical inference*, which uses principles of sampling theory.

One important problem of statistical inference is the estimation of *population parameters*, or briefly *parameters* (such as population mean and variance), from the corresponding *sample statistics*, or briefly *statistics* (such as sample mean and variance). We consider this problem in this chapter.

Unbiased Estimates

If the mean of the sampling distribution of a statistic equals the corresponding population parameter, the statistic is called an *unbiased estimator* of the parameter; otherwise, it is called a *biased estimator*. The corresponding values of such statistics are called *unbiased* or *biased estimates*, respectively.

EXAMPLE 6.1. The mean of the sampling distribution of means $\mu_{\bar{x}}$ is μ, the population mean. Hence the sample mean X is an unbiased estimate of the population mean μ.

EXAMPLE 6.2. The mean of the sampling distribution of variances is

$$\mu_{s^2} = \frac{N-1}{N} \sigma^2$$

where σ^2 is the population variance and N is the sample size (see Table 5-1). Thus the sample variance s^2 is a biased estimate of the population variance σ^2. By using the modified variance

$$\hat{s}^2 = \frac{N}{N-1} s^2$$

we find $\mu_\hat{s} = \sigma^2$, so that $\hat{s}^2$ is an unbiased estimate of σ^2. However, $\hat{s}$ is a biased estimate of σ.

In the language of expectation we could say that a statistic is unbiased if its expectation equals the corresponding population parameter. Thus X and $\hat{s}^2$ are unbiased since $E\{X\} = \mu$ and $E\{\hat{s}^2\} = \sigma^2$.

Efficient Estimates

If the sampling distributions of two statistics have the same mean (or expectation), then the statistic with the smaller variance is called an *efficient estimator* of the mean, while the other statistic is called an *inefficient estimator*. The corresponding values of the statistics are called *efficient estimates*, respectively.

Remember

If we consider all possible statistics whose sampling distributions have the same mean, the one with the smallest variance is sometimes called the *most efficient*, or *best, estimator* of this mean.

EXAMPLE 6.3. The sampling distributions of the mean and median both have the same mean, namely, the population mean. However, the variance of the sampling distribution of means is smaller than the variance of the sampling distribution of medians. Hence the sample mean gives an efficient estimate of the population mean, while the sample median gives an inefficient estimate of it.

Of all statistics estimating the population mean, the sample mean provides the best (or most efficient) estimate.

In practice, inefficient estimates are often used because of the relative ease with which some of them can be obtained.

Point Estimates and Interval Estimates

An estimate of a population parameter given by a single number is called a *point estimate* of the parameter. An estimate of a population parameter given by two numbers between which the parameter may be considered to lie is called an *interval estimate* of the parameter.

Interval estimates indicate the precision, or accuracy, of an estimate and are therefore preferable to point estimates.

EXAMPLE 6.4. If we say that a distance is measured as 5.28 meters (m), we are giving a point estimate. If, on the other hand, we say that the distance is 5.28 ± 0.03 m (i.e., the distance lies between 5.25 and 5.31 m), we are giving an interval estimate.

A statement of the error (or precision) of an estimate is often called its *reliability*.

Confidence-Interval Estimates of Population Parameters

Let μ_S and σ_S be the mean and standard deviation (standard error), respectively, of the sampling distribution of a statistic S. Then if the sampling distribution of S is approximately normal (which as we have seen is true for many statistics if the sample size $N \geq 30$), we can expect to find an actual sample statistic S lying in the intervals $\mu_S - \sigma_S$ to $\mu_S + \sigma_S$, $\mu_S - 2\sigma_S$ to $\mu_S + 2\sigma_S$, or $\mu_S - 3\sigma_S$ to $\mu_S + 3\sigma_S$ about 68.27%, 95.45%, and 99.73% of the time, respectively.

Equivalently, we can expect to find (or we can be *confident* of finding) μ_S in the intervals $S - \sigma_S$ to $S + \sigma_S$, $S - 2\sigma_S$ to $S + 2\sigma_S$, or $S - 3\sigma_S$ to $S + 3\sigma_S$ about 68.27%, 95.45%, and 99.73% of the time, respectively. Because of this, we call these respective intervals the 68.27%, 95.45%, and 99.73% *confidence intervals* for estimating σ_S. The end numbers of these intervals ($S \pm \sigma_S$, $S \pm 2\sigma_S$, and $S \pm 3\sigma_S$) are then called the 68.27%, 95.45%, and 99.73% *confidence limits*.

Similarly, $S \pm 1.96\sigma_S$, and $S \pm 2.58\sigma_S$ are, respectively, the 95% and 99% (or 0.95 and 0.99) confidence limits for S. The percentage confidence is often called the *confidence level*. The numbers 1.96, 2.58, etc., in the confidence limits are called *confidence coefficients*, or *critical values*, and are denoted by z_c. From confidence levels we can find confidence coefficients, and vice versa.

Table 6-1 shows the values of z_c corresponding to various confidence levels used in practice. For confidence levels not presented in the table, the values of z_c can be found from the normal-curve area tables (see Appendix A).

Table 6-1

Confidence Level	99.73%	99%	98%	96%	95.45%	95%	90%	80%	68.27%	50%
z_c	3.00	2.58	2.33	2.05	2.000	1.96	1.645	1.28	1.00	0.6745

Confidence Intervals for Means

If the statistic S is the sample mean , then the 95% and 99% confidence limits for estimating the population mean μ are given by $X \pm 1.96\sigma_{\bar{x}}$ and $X \pm 2.58\sigma_{\bar{x}}$, respectively. More generally, the confidence limits are given by $X \pm z_c\sigma_{\bar{x}}$, where z_c (which depends on the particular level of confidence desired) can be read from Table 6-1. Using the values of $\sigma_{\bar{x}}$, obtained in Chapter 5, we see that the confidence limits for the population mean are given by

$$\bar{X} \pm z_c \frac{\sigma}{\sqrt{N}} \tag{1}$$

if the sampling is either from an infinite population or with replacement from a finite population and are given by

$$\bar{X} \pm z_c \frac{\sigma}{\sqrt{N}} \sqrt{\frac{N_p - N}{N_p - 1}} \tag{2}$$

if the sampling is without replacement from a population of finite size N_p.

Generally, the population standard deviation σ is unknown; thus, to obtain the above confidence limits, we use the sample estimate $\hat{s}$ or s. This will prove satisfactory when $N \geq 30$. For $N < 30$, the approximation is poor and small sampling theory must be employed.

Confidence Intervals for Proportions

If the statistic S is the proportion of "successes" in a sample of size N drawn from a binomial population in which p is the proportion of successes (i.e., the probability of success), then the confidence limits for p are given by $P \pm z_c \sigma_p$, where P is the proportion of successes in the sample of size N. Using the values of σ_p obtained in Chapter 5, we see that the confidence limits for the population proportion are given by

$$P \pm z_c \sqrt{\frac{pq}{N}} = P \pm z_c \sqrt{\frac{p(1-p)}{N}} \tag{3}$$

if the sampling is either from an infinite population or with replacement from a finite population and are given by

$$P \pm z_c \sqrt{\frac{pq}{N}} \sqrt{\frac{N_p - N}{N_p - 1}} \tag{4}$$

if the sampling is without replacement from a population of finite size N_p.

To compute these confidence limits, we can use the sample estimate P for p, which will generally prove satisfactory if $N \geq 30$.

Confidence Intervals for Differences and Sums

If S_1 and S_2 are two sample statistics with approximately normal sampling distributions, confidence limits for the difference of the population parameters corresponding to S_1 and S_2 are given by

$$S_1 - S_2 \pm z_c \sigma_{S1-S2} = S_1 - S_2 \pm z_c \sqrt{\sigma_{S1}^2 + \sigma_{S2}^2} \tag{5}$$

while confidence limits for the sum of the population parameters are given by

$$S_1 + S_2 \pm z_c \sigma_{S1+S2} = S_1 + S_2 \pm z_c \sqrt{\sigma_{S1}^2 + \sigma_{S2}^2} \tag{6}$$

provided that the samples are independent.

For example, confidence limits for the difference of two population means, in the case where the populations are infinite, are given by

$$\bar{X}_1 - \bar{X}_2 \pm z_c\sigma_{\bar{X}1-\bar{X}2} = \bar{X}_1 - \bar{X}_2 \pm z_c \sqrt{\frac{\sigma_1^2}{N_1} + \frac{\sigma_2^2}{N_2}} \tag{7}$$

where X_1, σ_1, N_1 and X_2, σ_2, N_2 are the respective means, standard deviations, and sizes of the two samples drawn from the populations.

Similarly, confidence limits for the difference of two population proportions, where the populations are infinite, are given by

$$P_1 - P_2 \pm z_c\sigma_{P1-P2} = P_1 - P_2 \pm z_c \sqrt{\frac{p_1(1-p_1)}{N_1} + \frac{p_2(1-p_2)}{N_2}} \tag{8}$$

where P_1 and P_2 are the two sample proportions, N_1 and N_2 are the sizes of the two samples drawn from the populations, and p_1 and p_2 are the proportions in the two populations (estimated by P_1 and P_2).

Confidence Intervals for Standard Deviations

The confidence limits for the standard deviation σ of a normally distributed population, as estimated from a sample with standard deviation s, are given by

$$s \pm z_c\sigma_s = s \pm z_c \frac{\sigma}{\sqrt{2N}} \tag{9}$$

using Table 5-1. In computing these confidence limits, we use s or $\hat{s}$ to estimate σ.

Probable Error

The 50% confidence limits of the population parameters corresponding to a statistic S are given by $S \pm 0.6745\sigma_S$. The quantity $0.6745\sigma_S$ is known as the *probable error* of the estimate.

Chapter 7
STATISTICAL
DECISION THEORY

IN THIS CHAPTER:

✔ *Statistical Decisions and Hypotheses*
✔ *Tests of Hypotheses and Significance*
✔ *Type I and Type II Errors*
✔ *Level of Significance*
✔ *Tests Involving Normal Distributions*
✔ *Two-Tailed and One-Tailed Tests*
✔ *Special Tests*
✔ *Tests Involving Sample Differences*

Statistical Decisions and Hypotheses

Very often in practice we are called upon to make decisions about populations on the basis of sample information. Such decisions are called *statistical decisions*. For example, we may wish to decide on the basis of sample data whether a new serum is really effective in curing a disease, whether one educational procedure is better than another, or whether a given coin is loaded.

In attempting to reach decisions, it is useful to make assumptions (or guesses) about the populations involved. Such assumptions, which may or may not be true, are called *statistical hypotheses*. They are generally statements about the probability distributions of the populations.

Null Hypotheses

In many instances we formulate a statistical hypothesis to determine whether an expected characteristic or outcome is likely to be true. For example, if we want to decide whether a given coin is loaded, we formulate the hypothesis that the coin is fair (i.e., $p = 0.5$, where p is the probability of heads). Similarly, if we want to decide whether one procedure is better than another, we formulate the hypothesis that there is *no difference* between the procedures (i.e., any observed differences are due merely to fluctuations in sampling from the *same* population). Such hypotheses are often called *null hypotheses* and are denoted by H_0. Null hypotheses usually correspond to no difference in the characteristic or outcome of interest, or zero effect.

Alternative Hypotheses

Any hypothesis that differs from a given hypothesis is called an *alternative hypothesis*. For example, if one hypothesis is $p = 0.5$, alternative hypotheses might be $p = 0.7$, $p \neq 0.5$, or $p > 0.5$. A hypothesis alternative to the null hypothesis is denoted by H_1.

Tests of Hypotheses and Significance

If we suppose that a particular hypothesis is true but find that the results observed in a random sample differ markedly from the results expected under the hypothesis (i.e., expected on the basis of pure chance, using sampling theory), then we would say that the observed differences are *significant* and would thus be inclined to reject the hypothesis (or at least not accept it on the basis of the evidence obtained). For example, if 20 tosses of a coin yield 16 heads, we would be inclined to reject the hypothesis that the coin is fair, although it is conceivable that we might be wrong.

You Need to Know

Procedures that enable us to determine whether observed samples differ significantly from the results expected, and thus help us decide whether to accept or reject hypotheses, are called *tests of hypotheses, tests of significance, rules of decision*, or simply *decision rules*.

Type I and Type II Errors

If we reject a hypothesis when it should be accepted, we say that a *Type I error* has been made. If, on the other hand, we accept a hypothesis when it should be rejected, we say that a *Type II error* has been made. In either case, a wrong decision or error in judgment has occurred.

In order for decision rules (or tests of hypotheses) to be good, they must be designed so as to minimize errors of decision. This is not a simple matter, because for any given sample size, an attempt to decrease one type of error is generally accompanied by an increase in the other type of error. In practice, one type of error may be more serious than the other, and so a compromise should be reached in favor of limiting the more serious error. The only way to reduce both types of error is to increase the sample size, which may or may not be possible.

Level of Significance

In testing a given hypothesis, the maximum probability with which we would be willing to risk a Type I error is called the *level of significance*, or *significance level*, of the test. This probability, often denoted by α, is generally specified before any samples are drawn so that the results obtained will not influence our choice.

In practice, a significance level of 0.05 or 0.01 is customary, although other values are used. If, for example, the 0.05 (or 5%) significance level is chosen in designing a decision rule, then there are about 5 chances in 100 that we would reject the hypothesis when it should be accepted; that is, we are about 95% *confident* that we have made the right decision. In such case we say that the hypothesis has been rejected at the 0.05 significance level, which means that the hypothesis has a 0.05 probability of being wrong.

Tests Involving Normal Distributions

To illustrate the ideas presented above, suppose that under a given hypothesis the sampling distribution of a statistic S is a normal distribution with mean μ_S and standard deviation σ_S. Thus the distribution of the standardized variable (or z score), given by $z = (S - \mu_S) / \sigma_S$, is the standardized normal distribution (mean 0, variance 1), as shown in Fig. 7-1.

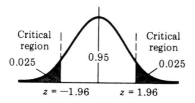

Figure 7-1

As indicated in Fig. 7-1, we can be 95% confident that if the hypothesis is true, then the z score of an actual sample statistic S will lie between -1.96 and 1.96 (since the area under the normal curve between these values is 0.95). However, if on choosing a single sample at random we find that the z score of its statistic lies *outside* the range -1.96 to 1.96, we would conclude that such an event could happen with a probability of only 0.05 (the total shaded area in the figure) if the given hypothesis were true. We would then say that this z score differed *significantly* from what would be expected under the hypothesis, and we would then be inclined to reject the hypothesis.

The total shaded area 0.05 is the significance level of the test. It represents the probability of our being wrong in rejecting the hypothesis (i.e., the probability of making a Type I error). Thus we say that the hypothesis is *rejected at the 0.05 significance level* or that the z score of the given sample statistic is *significant at the 0.05 level*.

The set of z scores outside the range -1.96 to 1.96 constitutes what is called *the critical region of the hypothesis, the region of rejection of the hypothesis*, or the *region of significance*. The set of z scores inside the range -1.96 to 1.96 is thus called the *region of acceptance of the hypothesis*, or the *region of nonsignificance*.

On the basis of the above remarks, we can formulate the following decision rule (or test of hypothesis or significance):

Reject the hypothesis at the 0.05 significance level if the z score of the statistic S lies outside the range -1.96 to 1.96 (i.e., either $z > 1.96$ or $z < -1.96$). This is equivalent to saying that the observed sample statistic is significant at the 0.05 level.

Accept the hypothesis otherwise (or, if desired, make no decision at all).

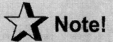 **Note!**

Because the z score plays such an important part in tests of hypotheses, it is also called a *test statistic*.

It should be noted that other significance levels could have been used. For example, if the 0.01 level were used, we would replace 1.96 everywhere above with 2.58 (see Table 7-1). Table 6-1 can also be used, since the sum of the significance and confidence levels is 100%.

Two-Tailed and One-Tailed Tests

In the above test we were interested in extreme values of the statistic S or its corresponding z score on *both* sides of the mean (i.e., in both tails of the distribution). Such tests are thus called *two-sided tests*, or *two-tailed tests*.

Often, however, we may be interested only in extreme values to one side of the mean (i.e., in one tail of the distribution), such as when we are testing the hypothesis that one process is better than another (which is different from testing whether one process is better or worse than the other). Such tests are called *one-sided tests*, or *one-tailed tests*. In such cases the critical region is a region to one side of the distribution, with an area equal to the level of significance.

Table 7-1, which gives critical values of z for both one-tailed and two- tailed tests at various levels of significance, will be found useful for reference purposes. Critical values of z for other levels of significance are found from the table of normal-curve areas (Appendix A).

Table 7-1

Level of significance, α	0.10	0.05	0.01	0.005	0.002
Critical values of z for one-tailed tests	-1.28 *or* 1.28	-1.645 *or* 1.645	-2.33 *or* 2.33	-2.58 *or* 2.58	-2.88 *or* 2.88
Critical values of z for two-tailed tests	-1.645 *and* 1.645	-1.96 *and* 1.96	-2.58 *and* 2.58	-2.81 *and* 2.81	-3.08 *and* 3.08

The area under the curve corresponding to the tails of the distribution defined by the z score of an actual sample statistic S is commonly called the *p-value* of the z score. The *p-value* is interpreted as the probability of observing an absolute z score greater than or equal to the absolute value actually observed if the null hypothesis were true. For example, in a two-tailed test, if the observed z score is 2.5, then the corresponding *p-value* is $2(0.5000 - .4938) = 0.0124$. The probability of observing an absolute z score greater than or equal to 2.5 if the null hypothesis were true is only 0.0124.

Special Tests

For large samples, the sampling distributions of many statistics are normal distributions (or at least nearly normal), and the above tests can be applied to the corresponding z scores. The following special cases, taken from Table 5-1, are just a few of the statistics of practical interest. In each case the results hold for infinite populations or for sampling with replacement. For sampling without replacement from finite populations, the results must be modified.

1. **Means**. Here $S = X$, the sample mean; $\mu_S = \mu_{\bar{x}} = \mu$, the population mean; and $\sigma_S = \sigma_{\bar{x}} = \sigma / \sqrt{N}$, where σ is the population standard deviation and N is the sample size. The z score is given by

$$z = \frac{X - \mu}{\sigma / \sqrt{N}}$$

When necessary, the sample deviation s or $\hat{s}$ is used to estimate σ.

2. **Proportions**. Here $S = P$, the proportion of "successes" in a sample; $\mu_S = \mu_P = p$, where p is the population proportion of successes and N is the sample size; and $\sigma_S = \sigma_P = \sqrt{pq / N}$, where $q = 1 - p$.

The z score is given by

$$z = \frac{P - p}{\sqrt{pq / N}}$$

In case $P = X / N$, where X is the actual number of successes in a sample, the z score becomes

$$z = \frac{X - Np}{\sqrt{Npq}}$$

That is, $\mu_X = \mu = Np$, $\sigma_X = \sigma = \sqrt{Npq}$, and $S = X$.

The results for other statistics can be obtained similarly.

Tests Involving Sample Differences

Differences of Means

Let X_1 and X_2 be the sample means obtained in large samples of sizes N_1 and N_2 drawn from respective populations having means μ_1 and μ_2 and standard deviations σ_1 and σ_2. Consider the null hypothesis that there is *no difference* between the population means (i.e., $\mu_1 = \mu_2$), which is to say that the samples are drawn from two populations having the same mean.

Placing μ_1 and μ_2 in equation (5) of Chapter 5, we see that the sampling distribution of differences in means is approximately normally distributed, with its mean and standard deviation given by

$$\mu_{\bar{X}_1 - \bar{X}_2} = 0 \quad \text{and} \quad \sigma_{\bar{X}_1 - \bar{X}_2} = \sqrt{\frac{\sigma_1^2}{N_1} + \frac{\sigma_2^2}{N_2}} \tag{1}$$

where we can, if necessary, use the sample standard deviations s_1 and s_2 (or $\hat{s}_1$ and $\hat{s}_2$) as estimates of σ_1 and σ_2.

By using the standardized variable, or z score, given by

$$z = \frac{\bar{X}_1 - \bar{X}_2 - 0}{\sigma_{\bar{X}_1 - \bar{X}_2}} = \frac{\bar{X}_1 - \bar{X}_2}{\sigma_{\bar{X}_1 - \bar{X}_2}} \tag{2}$$

we can test the null hypothesis against alternative hypotheses (or the significance of an observed difference) at an appropriate level of significance.

Differences of Proportions

Let P_1 and P_2 be the sample proportions obtained in large samples of sizes N_1 and N_2 drawn from respective populations having proportions p_1 and p_2. Consider the null hypothesis that there is *no difference* between the population parameters (i.e., $p_1 = p_2$) and thus that the samples are really drawn from the same population.

Placing $p_1 = p_2 = p$ in equation (6) of Chapter 5, we see that the sampling distribution of differences in proportions is approximately normally distributed, with its mean and standard deviation given by

$$\mu_{P1-P2} = 0 \quad \text{and} \quad \sigma_{P1-P2} = \sqrt{pq\left(\frac{1}{N_1} + \frac{1}{N_2}\right)}$$

$$(3)$$

where

$$p = \frac{N_1 P_1 + N_2 P_2}{N_1 + N_2}$$

is used as an estimate of the population proportion and where $q = 1 - p$. By using the standardized variable

$$z = \frac{P_1 - P_2 - 0}{\sigma_{P1-P2}} = \frac{P_1 - P_2}{\sigma_{P1-P2}}$$

$$(4)$$

we can test observed differences at an appropriate level of significance and thereby test the null hypothesis.

Tests involving other statistics can be designed similarly.

Chapter 8
SMALL SAMPLING
THEORY

IN THIS CHAPTER:

✔ *Small Samples*
✔ *Student's t Distribution*
✔ *Confidence Intervals*
✔ *Tests of Hypotheses*
  and Significance
✔ *The Chi-Square Distribution*
✔ *Confidence Intervals for χ^2*
✔ *Degrees of Freedom*
✔ *The F Distribution*

Small Samples

In previous chapters we often made use of the fact that for samples of size $N > 30$, called *large samples*, the sampling distributions of many statistics are approximately normal, the approximation becoming better with increasing N. For samples of size $N < 30$, called *small samples*, this approximation is not good and becomes worse with decreasing N, so that appropriate modifications must be made.

A study of sampling distributions of statistics for small samples is called *small sampling theory*. However, a more suitable name would be *exact sampling theory*, since the results obtained hold for large as well as for small samples. In this chapter we study three important distributions: Student's *t* distribution, the chi-square distribution, and the *F* distribution.

Student's *t* Distribution

Let us define the statistic

$$t = \frac{\bar{X} - \mu}{s}\sqrt{N-1} = \frac{\bar{X} - \mu}{\hat{s}/\sqrt{N}} \qquad (1)$$

which is analogous to the z statistic given by

$$z = \frac{\bar{X} - \mu}{\sigma/\sqrt{N}}$$

If we consider samples of size N drawn from a normal (or approximately normal) population with mean μ and if for each sample we compute t, using the sample mean X and sample standard deviation s or $\hat{s}$, the sampling distribution for t can be obtained and is called *Student's t distribution* after its discoverer, W. S. Gossett, who published his works under the pseudonym "Student" during the early part of the twentieth century. Student's *t* distribution is defined by three quantities, or parameters, the mean, variance, and the constant υ which equals $(N - 1)$ and is called the *number of degrees of freedom* (υ is the Greek letter *nu*). Figure 8.1 shows Student's *t* distribution for various values of υ. Like the standardized normal curve, the total area under the *t* curve is 1.

For large values of υ or N (certainly $N \geq 30$) the *t* curve closely approximates the standardized normal curve

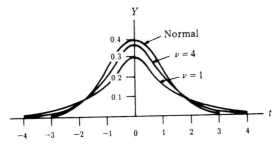

Figure 8-1

Confidence Intervals

As done with normal distributions in Chapter 6, we can define 95%, 99%, or other confidence intervals by using the table of the t distribution in Appendix B. In this manner we can estimate within specified limits of confidence the population mean μ.

For example, if $-t_{.975}$ and $t_{.975}$ are the values of t for which 2.5% of the area lies in each tail of the t distribution, then the 95% confidence interval for t is

$$-t_{.975} < \frac{\bar{X} - \mu}{s} \sqrt{N-1} < t_{.975} \qquad (2)$$

from which we see that μ is estimated to lie in the interval

$$\bar{X} - t_{.975} \frac{s}{\sqrt{N-1}} < \mu < \bar{X} + t_{.975} \frac{s}{\sqrt{N-1}} \qquad (3)$$

with 95% confidence (i.e. probability 0.95). Note that $t_{.975}$ represents the 97.5 percentile value, while $t_{.025} = -t_{.975}$ represents the 2.5 percentile value.

In general, we can represent confidence limits for population means by

$$\bar{X} \pm t_c \frac{s}{\sqrt{N-1}} \qquad (4)$$

where the values $\pm t_c$, called *critical values* or *confidence coefficients*, depend on the level of confidence desired and on the sample size. They can be read from the table in Appendix B.

A comparison of equation (4) with the confidence limits ($\bar{X} \ z_c \sigma$ / $\sqrt{N}$) of Chapter 6, shows that for small samples we replace z_c (obtained from the normal distribution) with t_c (obtained from the t distribution) and that we replace σ with $\sqrt{N} / (N - 1)s = \hat{s}$, which is the sample estimate of σ. As N increases, both methods tend toward agreement.

Tests of Hypotheses and Significance

Tests of hypotheses and significance, or decision rules (as discussed in Chapter 7), are easily extended to problems involving small samples, the only difference being that the z *score*, or z *statistic*, is replaced by a *suitable t score*, or t *statistic*.

1. **Means.** To test the hypothesis H_0 that a normal population has mean μ, we use the t score (or t statistic)

$$\bar{X} \pm t_c \frac{s}{\sqrt{N-1}} \tag{5}$$

where $\bar{X}$ is the mean of a sample of size N. This is analogous to using the z score

$$\sqrt{N/(N-1)}s = \hat{s},$$

for large N, except that $\hat{s} = \sqrt{N / (N - 1)}s$ is used in place of σ. The difference is that while z is normally distributed, t follows Student's distribution. As N increases, these tend toward agreement.

2. **Differences of Means.** Suppose that two random samples of sizes N_1 and N_2 are drawn from normal populations whose standard deviations are equal ($\sigma_1 = \sigma_2$). Suppose further that these two samples have means given by X_1 and X_2 and standard deviations given by s_1 and s_2, respectively. To test the hypothesis H_0 that the samples come from the same population (i.e., $\mu_1 = \mu_2$ as well as $\sigma_1 = \sigma_2$), we use the t score given by

$$t = \frac{\bar{X}_1 - \bar{X}_2}{\sigma\sqrt{1/N_1 + 1/N_2}} \qquad \text{where} \qquad \sigma = \sqrt{\frac{N_1 s_1^2 + N_2 s_2^2}{N_1 + N_2 - 2}} \qquad (6)$$

The distribution of t is Student's distribution with $v = N_1 + N_2 - 2$ degrees of freedom. The use of equation (6) is made plausible on placing $\sigma_1 = \sigma_2 = \sigma$ in the z score of equation (2) of Chapter 7 and then using as an estimate of σ^2 the weighted mean

$$\frac{(N_1 - 1)\hat{s}_1^2 + (N_2 - 1)\hat{s}_2^2}{(N_1 - 1) + (N_2 - 1)} = \frac{N_1 s_1^2 + N_2 s_2^2}{N_1 + N_2 - 2}$$

where $\hat{s}_1^2$ and $\hat{s}_2^2$ are the unbiased estimates σ_1^2 and σ_2^2.

The Chi-Square Distribution

Let us define the statistic

$$\chi^2 = \frac{N s^2}{\sigma^2} = \frac{(X_1 - \bar{X})^2 + (X_2 - \bar{X})^2 + \cdots + (X_N - \bar{X})^2}{\sigma^2} \qquad (7)$$

where χ is the Greek letter *chi* and χ^2 is read "chi-square."

If we consider samples of size N drawn from a normal population with standard deviation σ, and if for each sample we compute χ^2, a sampling distribution for χ^2, called the *chi-square distribution*, can be obtained. The chi-square distributions corresponding to various values of v are shown in Fig. 8-2, where $v = N - 1$ is the *number of degrees of freedom*, and the area under each curve is 1.

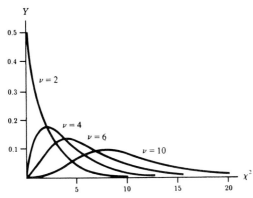

Figure 8-2 Chi-square distributions for various values of v

Confidence Intervals for χ^2

As done with the normal and t distribution, we can define 95%, 99%, or other confidence limits and intervals χ^2 by using the table of the χ^2 distribution in Appendix C. In this manner we can estimate within specified limits of confidence the population standard deviation σ, in terms of a sample standard deviation s.

For example, if $\chi^2_{.025}$ and $\chi^2_{.975}$ are the values (called *critical values*) for which 2.5% of the area lies in each tail of the distribution, then the 95% confidence interval is

$$\chi^2_{.025} < \frac{Ns^2}{\sigma^2} < \chi^2_{.975} \tag{8}.$$

from which we see that σ is estimated to lie in the interval

$$\frac{s\sqrt{N}}{\chi_{.975}} < \sigma < \frac{s\sqrt{N}}{\chi_{.025}} \tag{9}$$

with 95% confidence. Other confidence intervals can be found similarly. The values $\chi^2_{.025}$ and $\chi^2_{.975}$ represent, respectively, the 2.5 and 97.5 percentile values.

The table in Appendix C gives percentile values corresponding to the number of degrees of freedom υ. For large values of υ ($\upsilon \geq 30$), we can use the fact that $(\sqrt{2\chi^2} - \sqrt{2\upsilon - 1})$ is very nearly normally distributed with mean 0 and standard deviation 1; thus normal distribution tables can be used if $\upsilon \geq 30$. Then if χ^2_p and z_p are the pth percentiles of the chi-square and normal distributions, respectively, we have

$$\chi^2_p = \tfrac{1}{2}(z_p + \sqrt{2\nu - 1})^2 \tag{10}$$

In these cases, agreement is close to the results obtained in Chapters 5 and 6.

Degrees of Freedom

In order to compute a statistic such as (1) or (7), it is necessary to use observations obtained from a sample as well as certain population parameters. If these parameters are unknown, they must be estimated from the sample.

You Need to Know

The *number of degrees of freedom* of a statistic, generally denoted by v, is defined as the number N of independent observations in the sample (i.e., the sample size) minus the number k of population parameters, which must be estimated from sample observations. In symbols, $v = N - k$.

In the case of statistic (*1*), the number of independent observations in the sample is N, from which we can compute X and s. However, since we must estimate μ, $k = 1$ and so $v = N - 1$.

In the case of statistic (*7*), the number of independent observations in the sample is N, from which we can compute s. However, since we must estimate σ, $k = 1$ and so $v = N - 1$.

The F Distribution

As we have seen, it is important in some applications to know the sampling distribution of the difference in means $(S_1^2 - S_2^2)$ of two samples. Similarly, we may need the sampling distribution of the difference in variances $(S_1^2 - S_2^2)$. It turns out, however, that this distribution is rather complicated. Because of this, we consider instead the statistic S_1^2 / S_2^2, since a large or small ratio would indicate a large difference, while a ratio nearly equal to 1 would indicate a small difference. The sampling distribution in such a case can be found and is called the *F distribution*, named after R. A. Fisher.

More precisely, suppose that we have two samples, 1 and 2, of sizes N_1 and N_2, respectively, drawn from two normal (or nearly normal) populations having variances σ_1^2 and σ_2^2. Let us define the statistic

$$F = \frac{\hat{S}_1^2/\sigma_1^2}{\hat{S}_2^2/\sigma_2^2} = \frac{N_1 S_1^2/(N_1-1)\sigma_1^2}{N_2 S_2^2/(N_2-1)\sigma_2^2} \tag{11}$$

where

$$\hat{S}_1^2 = \frac{N_1 S_1^2}{N_1-1} \qquad \hat{S}_2^2 = \frac{N_2 S_2^2}{N_2-1} \tag{12}$$

Then the sampling distribution of F is called Fisher's F distribution, or briefly the F distribution, with $v_1 = N_1 - 1$ and $v_2 = N_2 - 1$ degrees of freedom. The curve has a shape similar to that shown in Fig. 8-3, although this shape can vary considerably for different values of v_1 and v_2.

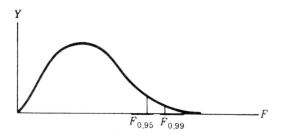

Figure 8-3

The F tables in Appendices D and E give percentile values of F for which the areas in the right-hand tail are 0.05 and 0.01, denoted by $F_{.95}$ and $F_{.99}$, respectively. Representing the 5% and 1% significance levels, these can be used to determine whether or not the variance S is significantly larger than S. In practice, the sample with the larger variance is chosen as sample 1.

Remember

While specially used with small samples, Student's t distribution, the chi-square distribution, and the F distribution are all valid for large sample sizes as well.

Chapter 9
THE CHI-SQUARE TEST

Observed and Theoretical Frequencies

As we have already seen many times, the results obtained in samples do not always agree exactly with the theoretical results expected according to the rules of probability. For example, although theoretical considerations lead us to expect 50 heads and 50 tails when we toss a fair coin 100 times, it is rare that these results are obtained exactly.

Suppose that in a particular sample a set of possible events E_1, E_2, E_3,..., E_k (see Table 9-1) are observed to occur with frequencies o_1, o_2, o_3,..., o_k called *observed frequencies*, and that according to probability rules they are expected to occur with frequencies e_1, e_2, e_3,..., e_k called *expected*, or *theoretical, frequencies*.

Table 9-1

Event	E_1	E_2	E_3	...	E_k
Observed frequency	o_1	o_2	o_3	...	o_k
Expected frequency	e_1	e_2	e_3	...	e_k

Often we wish to know whether the observed frequencies differ significantly from the expected frequencies. For the case where only two events E_1 and E_2 are possible (sometimes called a *dichotomy*, or *dichotomous classification*), as in the case of heads or tails, defective or nondefective bolts, etc., the problem is answered satisfactorily by the methods of previous chapters. In this chapter the general problem is considered.

Definition of χ^2

A measure of the discrepancy existing between the observed and expected frequencies is supplied by the statistic χ^2 (read chi-square) given by

$$\chi^2 = \frac{(o_1 - e_1)^2}{e_1} + \frac{(o_2 - e_2)^2}{e_2} + \cdots + \frac{(o_k - e_k)^2}{e_k} = \sum_{j=1}^{k} \frac{(o_j - e_j)^2}{e_j}$$

(1)

where if the total frequency is N,

$$\Sigma \, o_j = \Sigma \, e_j = N \qquad (2)$$

An expression equivalent to formula (*1*) is

$$\chi^2 = \Sigma \frac{o_j^2}{e_j} - N \qquad (3)$$

If $\chi^2 = 0$, the observed and theoretical frequencies agree exactly; while if $\chi^2 > 0$, they do not agree exactly. The larger the value of χ^2, the greater is the discrepancy between the observed and expected frequencies. The sampling distribution of χ^2 is approximated very closely by the chi-square distribution (already considered in Chapter 8) if the expected frequencies are at least equal to 5. The approximation improves for larger values.

The number of degrees of freedom, υ, is given by

1. $\upsilon = k - 1$ if the expected frequencies can be computed without having to estimate the population parameters from sample statistics. Note that we subtract 1 from k because of constraint condition (2), which states that if we know $k - 1$ of the expected frequencies, the remaining frequency can be determined.

2. $\upsilon = k - 1 - m$ if the expected frequencies can be computed only by estimating m population parameters from sample statistics.

Significance Tests

In practice, expected frequencies are computed on the basis of a hypothesis H_0. If under this hypothesis the computed value of χ^2 given by equation (*1*) or (*3*) is greater than some critical value (such as $\chi^2_{.95}$ or $\chi^2_{.99}$, which are the critical values of the 0.05 and 0.01 significance levels, respectively), we would conclude that the observed frequencies differ *significantly* from the expected frequencies and would reject H_0 at the corresponding level of significance; otherwise, we would accept it (or at least not reject it). This procedure is called the *chi-square test* of hypothesis or significance.

It should be noted that we must look with suspicion upon circumstances where χ^2 is *too close to zero*, since it is rare that observed frequencies agree *too well* with expected frequencies. To examine such situations, we can determine whether the computed value of χ^2 is less than $\chi^2_{.05}$ or $\chi^2_{.01}$, in which cases we would decide that the agreement is *too good* at the 0.05 or 0.01 significance levels, respectively.

The Chi-Square Test for Goodness of Fit

The chi-square test can be used to determine how well theoretical distributions (such as the normal and binomial distributions) fit empirical distributions (i.e., those obtained from sample data).

Contingency Tables

Table 9-1, in which the observed frequencies occupy a single row, is called a *one-way classification table*. Since the number of columns is k, this is also called a $1 \times k$ (read "1 by k") *table*. By extending these ideas, we can arrive at *two-way classification tables*, or $h \times k$ *tables*, in which the observed frequencies occupy h rows and k columns. Such tables are often called *contingency tables*. Contingency tables are constructed to determine whether there is a dependence between two categorical variables.

Corresponding to each observed frequency in an $h \times k$ contingency table, there is an *expected* (or *theoretical*) *frequency* that is computed subject to some hypothesis according to rules of probability. These frequencies, which occupy the *cells* of a contingency table, are called *cell frequencies*. The total frequency in each row or each column is called the *marginal frequency*.

To investigate agreement between the observed and expected frequencies, we compute the statistic

$$\chi^2 = \sum_j \frac{(o_j - e_j)^2}{e_j} \tag{4}$$

where the sum is taken over all cells in the contingency table and where the symbols o_j and e_j represent, respectively, the observed and expected frequencies in the jth cell. This sum, which is analogous to equation (1), contains hk terms. The sum of all observed frequencies is denoted by N and is equal to the sum of all expected frequencies [compare with equation (2)].

As before, statistic (4) has a sampling distribution given very closely by the chi-square distribution, provided the expected frequencies are not too small. The number of degrees of freedom, v, of this chi-square distribution is given for $h > 1$ and $k > 1$ by

1. $v = (h - 1)(k - 1)$ if the expected frequencies can be computed without having to estimate population parameters from sample statistics.

2. $v = (h - 1)(k - 1) - m$ if the expected frequencies can be computed only by estimating m population parameters from sample statistics.

Significance tests for $h \times k$ tables are similar to those for $1 \times k$ tables. The expected frequencies are found subject to a particular hypothesis H_0. A hypothesis commonly assumed is that the two classifications are independent of each other.

Contingency tables can be extended to higher dimensions. Thus, for example, we can have $h \times k \times l$ tables, where three classifications are present.

Yates' Correction for Continuity

When results for continuous distributions are applied to discrete data, certain corrections for continuity can be made, as we have seen in previous chapters. A similar correction is available when the chi-square distribution is used. The correction consists in rewriting equation (1) as

$$\chi^2 \text{ (corrected)} = \frac{(|o_1 - e_1| - 0.5)^2}{e_1} + \frac{(|o_2 - e_2| - 0.5)^2}{e_2} + \cdots + \frac{(|o_k - e_k| - 0.5)^2}{e_k} \qquad (5)$$

and is often referred to as *Yates' correction*. An analogous modification of equation (4) also exists.

In general, the correction is made only when the number of degrees of freedom is $v = 1$. For large samples, this yields practically the same results as the uncorrected χ^2, but difficulties can arise near critical values. For small samples where each expected frequency is between 5 and 10, it is perhaps best to compare both the corrected and uncorrected values of χ^2. If both values lead to the same conclusion regarding a hypothesis, such as rejection at the 0.05 level, difficulties are rarely encountered. If they lead to different conclusions, one can resort to increasing the sample sizes or, if this proves impractical, one can employ methods of probability involving the *multinomial distribution* of Chapter 4.

Simple Formulas for Computing χ^2

Simple formulas for computing χ^2 that involve only the observed frequencies can be derived. The following gives the results for 2×2 and 2×3 contingency tables (see Tables 9-2 and 9-3, respectively).

2×2 Tables

$$\chi^2 = \frac{N(a_1b_2 - a_2b_1)^2}{(a_1+b_1)(a_2+b_2)(a_1+a_2)(b_1+b_2)} = \frac{N\Delta^2}{N_1 N_2 N_A N_B} \tag{6}$$

Table 9-2

	I	II	Total
A	a_1	a_2	N_A
B	b_1	b_2	N_B
Total	N_1	N_2	N

Table 9-3

	I	II	III	Total
A	a_1	a_2	a_3	N_A
B	b_1	b_2	b_3	N_B
Total	N_1	N_2	N_3	N

where $\Delta = a_1b_2 - a_2b_1$, $N = a_1 + a_2 + b_1 + b_2$, $N_1 = a_1 + b_1$, $N_2 = a_2 + b_2$, $N_A = a_1 + a_2$, $N_B = b_1 + b_2$. With Yates' correction, this becomes

$$\chi^2 \text{(corrected)} = \frac{N(|a_1b_2 - a_2b_1| - \frac{1}{2}N)^2}{(a_1+b_1)(a_2+b_2)(a_1+a_2)(b_1+b_2)} = \frac{N(|\Delta| - \frac{1}{2}N)^2}{N_1 N_2 N_A N_B} \tag{7}$$

2 × 3 Tables

$$\chi^2 = \frac{N}{N_A}\left[\frac{a_1^2}{N_1} + \frac{a_2^2}{N_2} + \frac{a_3^2}{N_3}\right] + \frac{N}{N_B}\left[\frac{b_1^2}{N_1} + \frac{b_2^2}{N_2} + \frac{b_3^2}{N_3}\right] - N \tag{8}$$

where we have used the general result valid for all contingency tables:

$$\chi^2 = \sum \frac{o_j^2}{e_j} - N \tag{9}$$

Result (8) for $2 \times k$ tables where $k > 3$ can be generalized.

Additive Property of χ^2

Suppose that the results of repeated experiments yield sample values of χ^2 given by $\chi_1^2, \chi_2^2, \chi_3^2, \ldots$ with $v_1, v_2, v_3, \ldots$ degrees of freedom, respectively. Then the result of all these experiments can be considered equivalent to a χ^2 value given by $\chi_1^2 + \chi_2^2 + \chi_3^2 + \ldots$ with $v_1 + v_2 + v_3 + \ldots$ degrees of freedom.

Chapter 10
CURVE FITTING AND THE METHOD OF LEAST SQUARES

IN THIS CHAPTER:

✔ *Relationship between Variables*
✔ *Curve Fitting*
✔ *Equations of Approximating Curves*
✔ *The Straight Line*
✔ *The Method of Least Squares*
✔ *The Least-Squares Line*
✔ *The Least-Squares Parabola*
✔ *Regression*
✔ *Problems Involving More than Two Variables*

Relationship between Variables

Very often in practice a relationship is found to exist between two (or more) variables. For example, weights of adult males depend to some

76

degree on their heights, the circumferences of circles depend on their radii, and the pressure of a given mass of gas depends on its temperature and volume.

It is frequently desirable to express this relationship in mathematical form by determining an equation that connects the variables.

Curve Fitting

To determine an equation that connects variables, a first step is to collect data that show corresponding values of the variables under consideration. For example, suppose that X and Y denote, respectively, the height and weight of adult males; then a sample of N individuals would reveal the heights X_1, X_2,..., X_N and the corresponding weights Y_1, Y_2,..., Y_N.

A next step is to plot the points (X_1, Y_1), (X_2, Y_2),..., (X_N, Y_N) on a rectangular coordinate system. The resulting set of points is sometimes called a *scatter diagram*.

From the scatter diagram it is often possible to visualize a smooth curve that approximates the data. Such a curve is called an *approximating curve*. In Fig. 10-1, for example, the data appear to be approximated well by a straight line, and so we say that a *linear relationship* exists between the variables. In Fig. 10-2, however, although a relationship exists between the variables, it is not a linear relationship, and so we call it a *nonlinear relationship*.

The general problem of finding equations of approximating curves that fit given sets of data is called *curve fitting*.

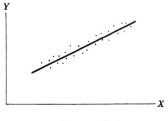

Figure 10-1

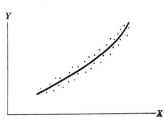

Figure 10-2

Equations of Approximating Curves

Several common types of approximating curves and their equations are listed below for reference purposes. All letters other than X and Y represent constants. The variables X and Y are often referred to as *independent* and *dependent variables*, respectively, although these roles can be interchanged.

Straight line	$Y = a_0 + a_1 X$	(1)
Parabola, or quadratic curve	$Y = a_0 + a_1 X + a_2 X^2$	(2)
Cubic curve	$Y = a_0 + a_1 X + a_2 X^2 + a_3 X^3$	(3)
Quartic curve	$Y = a_0 + a_1 X + a_2 X^2 + a_3 X^3 + a_4 X^4$	(4)
nth-Degree curve	$Y = a_0 + a_1 X + a_2 X^2 + \ldots + a_n X^n$	(5)

The right sides of the above equations are called *polynomials* of the first, second, third, fourth, and nth degrees, respectively. The functions defined by the first four equations are sometimes called *linear*, *quadratic*, *cubic*, and *quartic* functions, respectively.

The following are some of the many other equations frequently used in practice:

Hyperbola $\qquad Y = \dfrac{1}{a_0 + a_1 X} \qquad$ or $\qquad \dfrac{1}{Y} = a_0 + a_1 X \qquad$ (6)

Exponential curve $\qquad Y = ab^X \qquad$ (7)

$\qquad$ or $\qquad \log Y = \log a + (\log b) X = a_0 + a_1 X$

Geometric curve $\qquad Y = aX^b \qquad$ or $\qquad \log Y = \log a + b(\log X)$ (8)

Modified exponential curve $\qquad Y = ab^X + g \qquad$ (9)

Modified geometric curve $\qquad Y = aX^b + g \qquad$ (10)

Gompertz curve $\qquad Y = pq^{b^X} \qquad$ or $\qquad$ (11)

$$\log Y = \log p + b^X (\log q) = ab^X + g$$

Modified Gompertz curve	$Y = pq^{b^X} + h$	(12)

Logistic curve
$$Y = \frac{1}{ab^X + g} \quad \text{or} \quad \frac{1}{Y} = ab^X + g \quad (13)$$

$$Y = a_0 + a_1(\log X) + a_2(\log X)^2 \quad (14)$$

To decide which curve should be used, it is helpful to obtain scatter diagrams of transformed variables. For example, if a scatter diagram of log Y versus X shows a linear relationship, the equation has the form (7), while if log Y versus log X shows a linear relationship, the equation has the form (8).

The Straight Line

The simplest type of approximating curve is a straight line, whose equation can be written

$$Y = a_0 + a_1 X \quad (15)$$

Given any two points (X_1, Y_1) and (X_2, Y_2) on the line, the constants a_0 and a_1 can be determined. The resulting equation of the line can be written

$$Y = a_0 + a_1 X \quad (16)$$

where

$$Y - Y_1 = \left(\frac{Y_2 - Y_1}{X_2 - X_1}\right)(X - X_1) \quad \text{or} \quad Y - Y_1 = m(X - X_1)$$

$$m = \frac{Y_2 - Y_1}{X_2 - X_1}$$

is called the *slope* of the line and represents the change in Y divided by the corresponding change in X.

When the equation is written in the form (15), the constant a_1 is the slope m. The constant a_0, which is the value of Y when $X = 0$, is called the *Y intercept*.

The Method of Least Squares

To avoid individual judgment in constructing lines, parabolas, or other approximating curves to fit sets of data, it is necessary to agree on a definition of a "best-fitting line," "best-fitting parabola," etc.

By way of forming a definition, consider Fig. 10-3, in which the data points are given by $(X_1, Y_1), (X_2, Y_2),..., (X_N, Y_N)$. For a given value of X, say X_1, there will be a difference between the value Y_1 and the corresponding value as determined from the curve C. As shown in the figure, we denote this difference by D_1, which is sometimes referred to as a *deviation*, *error*, or *residual* and may be positive, negative, or zero. Similarly, corresponding to the values $X_2,..., X_N$ we obtain the deviations $D_2,..., D_N$.

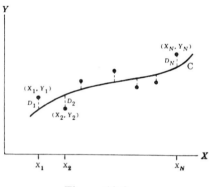

Figure 10-3

A measure of the "goodness of fit" of the curve C to the given data is provided by the quantity $D + D + ... + D$. If this is small, the fit is good; if it is large, the fit is bad. We therefore state the following:

Definition: Of all curves approximating a given set of data points, the curve having the property that $D + D + ... + D$ is a minimum is called a *best-fitting curve*.

A curve having this property is said to fit the data in the *least-squares sense* and is called a *least-squares curve*. Thus a line having this property is called a *least-squares line*, a parabola with this property is called a *least-squares parabola*, etc.

It is customary to employ the above definition when X is the independent variable and Y is the dependent variable. If X is the dependent variable, the definition is modified by considering horizontal instead of vertical deviations, which amounts to an interchange of the X and Y axes. These two definitions generally lead to different least-squares curves. Unless otherwise specified, we shall consider Y the dependent variable and X the independent variable.

Note!

It is possible to define another least-squares curve by considering perpendicular distances from each of the data points to the curve instead of either vertical or horizontal distances. However, this is not used very often.

The Least-Squares Line

The least-squares line approximating the set of points (X_1, Y_1), (X_2, Y_2),..., (X_N, Y_N) has the equation

$$Y = a_0 + a_1 X \qquad (17)$$

where the constants a_0 and a_1 are determined by solving simultaneously the equations

$$\sum Y = a_0 N + a_1 \sum X$$
$$\sum XY = a_0 \sum X + a_1 \sum X^2 \qquad (18)$$

which are called the *normal equations for the least-squares line (17)*. The constants a_0 and a_1 of equations (18) can, if desired, be found from the formulas

$$a_0 = \frac{(\sum Y)(\sum X^2) - (\sum X)(\sum XY)}{N \sum X^2 - (\sum X)^2} \qquad a_1 = \frac{N \sum XY - (\sum X)(\sum Y)}{N \sum X^2 - (\sum X)^2} \qquad (19)$$

The normal equations (18) are easily remembered by observing that the first equation can be obtained formally by summing on both sides of (17) [i.e., $\Sigma Y = \Sigma (a_0 + a_1 X) = a_0 N + a_1 \Sigma X$], while the second equation is obtained formally by first multiplying both sides of (17) by X and then summing [i.e., $\Sigma XY = \Sigma X(a_0 + a_1 X) = a_0 \Sigma X + a_1 \Sigma X^2$]. Note that this is not a derivation of the normal equations, but simply a means for remembering them. Note also that in equations (18) and (19) we have used the short notation ΣX, ΣXY, etc., in place of $\Sigma_{j=1}^{N} X_j$, $\Sigma_{j=1}^{N} X_j Y_j$, etc.

The labor involved in finding a least-squares line can sometimes be shortened by transforming the data so that $x = X - \overline{X}$ and $y = Y - \overline{Y}$. The equation of the least-squares line can then be written

$$y = \left(\frac{\Sigma xy}{\Sigma x^2}\right) x \quad \text{or} \quad y = \left(\frac{\Sigma xY}{\Sigma x^2}\right) x \tag{20}$$

In particular, if X is such that $\Sigma X = 0$ (i.e., $\overline{X} = 0$), this becomes

$$Y = \overline{Y} + \left(\frac{\Sigma XY}{\Sigma X^2}\right) X \tag{21}$$

Equation (20) implies that $y = 0$ when $x = 0$; thus the least-squares line passes through the point $(\overline{X}, \overline{Y})$, called the *centroid*, or *center of gravity*, of the data.

If the variable X is taken to be the dependent instead of the independent variable, we write equation (17) as $X = b_0 + b_1 Y$. Then the above results hold if X and Y are interchanged and a_0 and a_1 are replaced by b_0 and b_1, respectively. The resulting least-squares line, however, is generally not the same as that obtained above.

The Least-Squares Parabola

The least-squares parabola approximating the set of points (X_1, Y_1), (X_2, Y_2),..., (X_N, Y_N) has the equation

$$Y = a_0 + a_1 X + a_2 X^2 \tag{22}$$

where the constants a_0, a_1, and a_2 are determined by solving simultaneously the equations

$$\begin{aligned}
\sum Y &= a_0 \sum N + a_1 \sum X + a_2 \sum X^2 \\
\sum XY &= a_0 \sum X + a_1 \sum X^2 + a_2 \sum X^3 \qquad (23) \\
\sum X^2 Y &= a_0 \sum X^2 + a_1 \sum X^3 + a_2 \sum X^4
\end{aligned}$$

called the *normal equations for the least-squares parabola* (22).

Equations (23) are easily remembered by observing that they can be obtained formally by multiplying equation (22) by 1, X, and X^2, respectively, and summing on both sides of the resulting equations. This technique can be extended to obtain normal equations for least-squares cubic curves, least-squares quartic curves, and in general any of the least-squares curves corresponding to equation (5).

As in the case of the least-squares line, simplifications of equations (23) occur if X is chosen so that $\sum X = 0$. Simplification also occurs by choosing the new variables $x = X - X$ and $y = Y - Y$.

Regression

Often, on the basis of sample data, we wish to estimate the value of a variable Y corresponding to a given value of a variable X. This can be accomplished by estimating the value of Y from a least-squares curve that fits the sample data. The resulting curve is called a *regression curve of Y on X* since Y is estimated from X.

If we wanted to estimate the value of X from a given value of Y, we would use a *regression curve of X on Y*, which amounts to interchanging the variables in the scatter diagram so that X is the dependent variable and Y is the independent variable.

Remember!

In general, the regression line or curve of Y on X is not the same as the regression line or curve of X on Y.

Problems Involving More Than Two Variables

Problems involving more than two variables can be treated in a manner analogous to that for two variables. For example, there may be a relationship between the three variables X, Y, and Z that can be described by the equation

$$Z = a_0 + a_1 X + a_2 Y \qquad (24)$$

which is called a *linear equation in the variables X, Y, and Z.*

In a three-dimensional rectangular coordinate system this equation represents a plane, and the actual sample points (X_1, Y_1, Z_1), (X_2, Y_2, Z_2),..., (X_N, Y_N, Z_N) may "scatter" not too far from this plane, which we call an *approximating plane.*

By extension of the method of least squares, we can speak of a *least- squares plane* approximating the data. If we are estimating Z from given values of X and Y, this would be called a *regression plane of Z on X and Y.* The normal equations corresponding to the least-squares plane (24) are given by

$$
\begin{aligned}
\Sigma Z &= a_0 N + a_1 \Sigma X + a_2 \Sigma Y \\
\Sigma XZ &= a_0 \Sigma X + a_1 \Sigma X^2 + a_2 \Sigma XY \\
\Sigma YZ &= a_0 \Sigma Y + a_1 \Sigma XY + a_2 \Sigma Y^2
\end{aligned} \qquad (25)
$$

and can be remembered as being obtained from equation (24) by multiplying by 1, X, and Y successively and then summing.

More complicated equations than (24) can also be considered. These represent *regression surfaces.* If the number of variables exceeds three, geometric intuition is lost since we then require four-, five-,... dimensional spaces.

Chapter 11
CORRELATION THEORY

IN THIS CHAPTER:

Correlation and Regression

In the last chapter we considered the problem of *regression*, or *estimation*, of one variable (the dependent variable) from one or more related variables (the independent variables). In this chapter we consider the closely related problem of *correlation*, or the degree of relationship between variables, which seeks to determine *how well* a linear or other equation describes or explains the relationship between variables.

If all values of the variables satisfy an equation exactly, we say that the variables are *perfectly correlated* or that there is *perfect correlation* between them. Thus the circumferences C and radii r of all circles are perfectly correlated since $C = 2\pi r$. If two dice are tossed simultaneously 100 times, there is no relationship between corresponding points on each die (unless the dice are loaded); that is, they are *uncorrelated*. Such variables as the height and weight of individuals would show some correlation.

When only two variables are involved, we speak of *simple correlation* and *simple regression*. When more than two variables are involved, we speak of *multiple correlation* and *multiple regression*. This chapter considers only simple correlation. Multiple correlation and regression are considered in Chapter 12.

Linear Correlation

If X and Y denote the two variables under consideration, a *scatter diagram* shows the location of points (X, Y) on a rectangular coordinate system. If all points in this scatter diagram seem to lie near a line, as in Figs. 11-1(a) and 11-1(b), the correlation is called linear. In such cases, as we have seen in Chapter 10, a linear equation is appropriate for purposes of regression (or estimation).

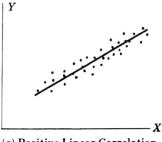

(*a*) **Positive Linear Correlation**

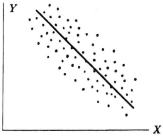

(*b*) **Negative Linear Correlation**

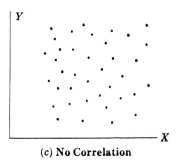

(*c*) **No Correlation**

Figure 11-1

If Y tends to increase as X increases, as in Fig. 11-1 (*a*), the correlation is called *positive*, or *direct*, *correlation*. If Y tends to decrease as
X increases, as in Fig. 11-1(*b*), the correlation is called *negative*, or
inverse, *correlation*.

If all points seem to lie near some curve, the correlation is called *nonlinear*, and a nonlinear equation is appropriate for regression, as we have seen in Chapter 10. It is clear that nonlinear correlation can be sometimes positive and sometimes negative.

If there is no relationship indicated between the variables, as in Fig. 11-1(*c*), we say that there is *no correlation* between them (i.e., they are *uncorrelated*).

Measures of Correlation

We can determine in a *qualitative* manner how well a given line or curve describes the relationship between variables by direct observation of the scatter diagram itself. For example, it is seen that a straight line is far more helpful in describing the relation between *X* and *Y* for the data of Fig. 11-1(*a*) than for the data of Fig. 11-1(*b*) because of the fact that there is less scattering about the line of Fig. 11-1(*a*).

If we are to deal with the problem of scattering of sample data about lines or curves in a *quantitative* manner, it will be necessary for us to devise *measures of correlation*.

The Least-Squares Regression Lines

We first consider the problem of how well a straight line explains the relationship between two variables. To do this, we shall need the equations for the least-squares regression lines obtained in Chapter 10. As we have seen, the least-squares regression line of *Y* on *X* is

$$Y = a_0 + a_1 X \qquad (1)$$

where a_0 and a_1 are obtained from the normal equations

$$\begin{aligned} \sum Y &= a_0 N + a_1 \sum X \\ \sum XY &= a_0 \sum X + a_1 \sum X^2 \end{aligned} \qquad (2)$$

which yield

$$a_0 = \frac{(\sum Y)(\sum X^2) - (\sum X)(\sum XY)}{N \sum X^2 - (\sum X)^2} \qquad (3)$$

$$a_1 = \frac{N \sum XY - (\sum X)(\sum Y)}{N \sum X^2 - (\sum X)^2}$$

Similarly, the regression line of X on Y is given by

$$X = b_0 + b_1 Y \qquad (4)$$

where b_0 and b_1 are obtained from the normal equations

$$\sum Y = b_0 N + b_1 \sum Y \qquad (5)$$
$$\sum XY = b_0 \sum X + b_1 \sum Y^2$$

which yield

$$b_0 = \frac{(\sum X)(\sum Y^2) - (\sum Y)(\sum XY)}{N \sum Y^2 - (\sum Y)^2}$$

$$b_1 = \frac{N \sum XY - (\sum X)(\sum Y)}{N \sum Y^2 - (\sum Y)^2} \qquad (6)$$

Equations (*1*) and (*4*) can also be written, respectively, as

$$y = \left(\frac{\sum xy}{\sum x^2} \right) x \qquad \text{and} \qquad x = \left(\frac{\sum xy}{\sum y^2} \right) y \qquad (7)$$

where $x = X - X$ and $y = Y - Y$.

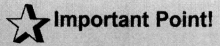

★ Important Point!

The regression equations are identical if and only if all points of the scatter diagram lie on a line. In such case there is *perfect linear correlation* between X and Y.

Standard Error of Estimate

If we let Y_{est} represent the value of Y for given values of X as estimated from equation (1), a measure of the scatter about the regression line of Y on X is supplied by the quantity

$$s_{Y.X} = \sqrt{\frac{\sum (Y - Y_{est})^2}{N}} \tag{8}$$

which is called the *standard error of estimate of Y on X*.

If the regression line (4) is used, an analogous standard error of estimate of X on Y is defined by

$$s_{X.Y} = \sqrt{\frac{\sum (X - X_{est})^2}{N}} \tag{9}$$

In general, $s_{Y.X} \neq s_{X.Y}$.

Equation (8) can be written

$$s_{Y.X}^2 = \frac{\sum Y^2 - a_0 \sum Y - a_1 \sum XY}{N} \tag{10}$$

which may be more suitable for computation. A similar expression exists for equation (9).

The standard error of estimate has properties analogous to those of the standard deviation. For example, if we construct lines parallel to the regression line of Y on X at respective vertical distances $s_{Y.X}$, $2s_{Y.X}$, $3s_{Y.X}$ from it, we should find, if N is large enough, that there would be included between these lines about 68%, 95%, and 99.7% of the sample points.

Just as a modified standard deviation given by

$$\hat{s} = \sqrt{\frac{N}{N-1}} \, s$$

was found useful for small samples, so a modified standard error of estimate given by

$$\hat{s}_{Y.X} = \sqrt{\frac{N}{N-2}} \, s_{Y.X}$$

is useful. For this reason, some statisticians prefer to define equation (8) or (9) with $N - 2$ replacing N in the denominator.

Explained and Unexplained Variation

The *total variation* of Y is defined as $\Sigma \, (Y - Y)^2$: that is, the sum of the squares of the deviations of the values of Y from the mean. This can be written

$$\Sigma \, (Y - Y)^2 = \Sigma \, (\bar{Y} - Y_{est})^2 + \Sigma \, (Y_{est} - \bar{Y})^2 \qquad (11)$$

The first term on the right of equation (11) is called the *unexplained variation*, while the second term is called the *explained variation* — so called because the deviations $Y_{est} - Y$ have a definite pattern, while the deviations $Y - Y_{est}$ behave in a random or unpredictable manner. Similar results hold for the variable X.

Coefficient of Correlation

The ratio of the explained variation to the total variation is called the *coefficient of determination*. If there is zero explained variation (i.e., the total variation is all unexplained), this ratio is 0. If there is zero unexplained variation (i.e., the total variation is all explained), the ratio is 1. In other cases the ratio lies between 0 and 1. Since the ratio is always nonnegative, we denote it by r^2. The quantity r, called the *coefficient of correlation* (or briefly *correlation coefficient*), is given by

$$r = \pm \sqrt{\frac{\text{explained variation}}{\text{total variation}}} = \pm \sqrt{\frac{\Sigma \, (Y_{est} - \bar{Y})^2}{\Sigma \, (Y - \bar{Y})^2}} \qquad (12)$$

and varies between -1 and +1. The + and - signs are used for positive linear correlation and negative linear correlation, respectively. Note that r is a dimension less quantity; that is, it does not depend on the units employed.

By using equations (*8*) and (*11*) and the fact that the standard deviation of Y is

$$s_Y = \sqrt{\frac{\sum(Y - \bar{Y})^2}{N}} \qquad (13)$$

we find that equation (*12*) can be written, disregarding the sign, as

$$r = \sqrt{1 - \frac{s_{Y.X}^2}{s_Y^2}} \qquad \text{or} \qquad s_{Y.X} = s_Y\sqrt{1 - r^2} \qquad (14)$$

Similar equations exist when X and Y are interchanged.

For the case of linear correlation, the quantity r is the same regardless of whether X or Y is considered the independent variable. Thus r is a very good measure of the linear correlation between two variables.

Concerning the Correlation Coefficient

The definitions of the correlation coefficient in equations (*12*) and (*14*) are quite general and can be used for nonlinear relationships as well as for linear ones, the only differences being that Y_{est} is computed from a nonlinear regression equation in place of a linear equation and that the + and - signs are omitted. In such case equation (*8*), defining the standard error of estimate, is perfectly general. Equation (*10*), however, which applies to linear regression only, must be modified. If, for example, the estimating equation is

$$Y = a_0 + a_1 X + a_2 X^2 + \cdots + a_{n-1} X^{n-1} \qquad (15)$$

then equation (*10*) is replaced by

$$s_{Y.X}^2 = \frac{\sum Y^2 - a_0 \sum Y - a_1 \sum XY - \cdots - a_{n-1} \sum X^{n-1} Y}{N} \qquad (16)$$

In such case the *modified standard error of estimate* (discussed earlier in this chapter) is

$$\hat{s}_{Y.X} = \sqrt{\frac{N}{N-n}} \, s_{Y.X}$$

where the quantity N - n is called the number of *degrees of freedom.*

It must be emphasized that in every case the computed value of r measures the degree of the relationship relative to the type of equation that is actually assumed. Thus if a linear equation is assumed and equation (*12*) or (*14*) yields a value of r near zero, it means that there is almost no *linear correlation* between the variables. However, it does not mean that there is no correlation at all, since there may actually be a high *nonlinear correlation* between the variables. In other words, the correlation coefficient measures the goodness of fit between (1) the equation actually assumed and (2) the data. Unless otherwise specified, the term *correlation coefficient* is used to mean *linear correlation coefficient.*

It should also be pointed out that a high correlation coefficient (i.e., near 1 or -1) does not necessarily indicate a direct dependence of the variables. For example there may be a high correlation between the number of books published each year and the number of thunderstorms each year even though the two events are clearly causally unrelated to one another. Such examples are sometimes referred to as *nonsense*, or *spurious*, *correlations*.

Product-Moment Formula for the Linear Correlation Coefficient

If a linear relationship between two variables is assumed, equation (*12*) becomes

$$r = \frac{\Sigma \, xy}{\sqrt{(\Sigma \, x^2)(\Sigma \, y^2)}} \tag{17}$$

where $x = X$ - $\bar{X}$ and $y = Y$ - $\bar{Y}$. This formula, which automatically gives the proper sign of r, is called the *product-moment formula* and clearly shows the symmetry beween X and Y.

If we write

$$s_{XY} = \frac{\Sigma\, xy}{N - 1} \qquad s_X = \sqrt{\frac{\Sigma\, x^2}{N - 1}} \qquad s_Y = \sqrt{\frac{\Sigma\, y^2}{N - 1}} \tag{18}$$

then s_X and s_Y will be recognized as the standard deviations of the variables X and Y, respectively, while s^2x and s^2y are their variances. The new quantity s_{XY} is called the *covariance* of X and Y. In terms of the symbols of formulas (*18*), formula (*17*) can be written

$$r = \frac{s_{XY}}{s_X s_Y} \tag{19}$$

Note that r is not only independent of the choice of units of X and Y, but is also independent of the choice of origin.

Sampling Theory of Correlation

The N pairs of values (X, Y) of two variables can be thought of as samples from a population of all such pairs that are possible. Since two variables are involved, this is called a *bivariate population*, which we assume to be a *bivariate normal distribution*.

We can think of a theoretical population coefficient of correlation, denoted by ρ, which is estimated by the sample correlation coefficient r. Tests of significance or hypotheses concerning various values of ρ require knowledge of the sampling distribution of r. For $\rho = 0$ this distribution is symmetrical, and a statistic involving Student's distribution can be used.

Test of Hypothesis $\rho = 0$. Here we use the fact that the statistic

$$t = \frac{r\sqrt{N - 2}}{\sqrt{1 - r^2}} \tag{20}$$

has Student's distribution with $\upsilon = N - 2$ degrees of freedom.

Sampling Theory of Regression

The regression equation $Y = a_0 + a_1 X$ is obtained on the basis of sample data. We are often interested in the corresponding regression equation for the population from which the sample was drawn. The following are three tests concerning such a population:

1. **Test of Hypothesis** $a_1 = A_1$. To test the hypothesis that the regression coefficient a_1 is equal to some specified value A_1, we use the fact that the statistic

$$t = \frac{a_1 - A_1}{s_{Y.X}/s_X}\sqrt{N-2} \tag{21}$$

has Student's distribution with N - 2 degrees of freedom. This can also be used to find confidence intervals for population regression coefficients from sample values.

2. **Test of Hypothesis for Predicted Values.** Let Y_0 denote the predicted value of Y corresponding to $X = X_0$ as estimated from the sample regression equation (i.e., $Y_0 = a_0 + a_1X_0$). Let $\overline{Y}_p$ denote the predicted value of Y corresponding to $X - X_0$ for the population. Then the statistic

$$t = \frac{Y_0 - Y_p}{s_{Y.X}\sqrt{N+1+(X_0-\bar{X})^2/s_X^2}}\sqrt{N-2} = \frac{Y_0 - Y_p}{\hat{s}_{Y.X}\sqrt{1+1/N+(X_0-\bar{X})^2/(Ns_X^2)}} \tag{22}$$

has Student's distribution with N - 2 degrees of freedom. From this, confidence limits for predicted population values can be found.

3. **Test of Hypothesis for Predicted Mean Values.** Let Y_0 denote the predicted value of Y corresponding to $X = X_0$ as estimated from the sample regression equation (i.e., $Y_0 = a_0 + a_1X_0$). Let $\overline{Y}_p$ denote the predicted *mean value* of Y corresponding to $X - X_0$ for the population. Then the statistic

$$t = \frac{Y_0 - \bar{Y}_p}{s_{Y.X}\sqrt{1+(X_0-\bar{X})^2/s_X^2}}\sqrt{N-2} = \frac{Y_0 - \bar{Y}_p}{\hat{s}_{Y.X}\sqrt{1/N+(X_0-\bar{X})^2/(Ns_X^2)}} \tag{23}$$

has Student's distribution with N - 2 degrees of freedom. From this, confidence limits for predicted mean population values can be found.

Chapter 12
MULTIPLE AND PARTIAL CORRELATION

IN THIS CHAPTER:

- ✔ *Multiple Correlation*
- ✔ *Subscript Notation*
- ✔ *Regression Equations and Regression Planes*
- ✔ *Normal Equations for the Least-Squares Regression Plane*
- ✔ *Regression Planes and Correlation Coefficients*
- ✔ *Standard Error of Estimate*
- ✔ *Coefficient of Multiple Correlation*
- ✔ *Change of Dependent Variable*
- ✔ *Generalizations to More than Three Variables*
- ✔ *Partial Correlation*

✔ Relationships between Multiple and Partial Correlation Coefficients

Multiple Correlation

The degree of relationship existing between three or more variables is called multiple correlation. The fundamental principles involved in problems of multiple correlation are analogous to those of simple correlation, as treated in Chapter 11.

Subscript Notation

To allow for generalizations to large numbers of variables, it is convenient to adopt a notation involving subscripts.

We shall let $X_1, X_2, X_3,...$ denote the variables under consideration. Then we can let $X_{11}, X_{12}, X_{13},...$ denote the values assumed by the variable X_1, and $X_{21}, X_{22}, X_{23},...$ denote the values assumed by the variable X_2, and so on. With this notation, a sum such as $X_{21} + X_{22} + X_{23} + ... + X_{2N}$ could be written $\sum_{j=1}^{N} X_{2j}, \sum_j X_{2j}$, or simply $\sum X_2$. When no ambiguity can result, we use the last notation. In such case the mean of X_2 is written $X_2 = (\sum X_2) / N$.

Regression Equations and Regression Planes

A *regression equation* is an equation for estimating a dependent variable, say X_1, from the independent variables $X_2, X_3,...$ and is called a *regression equation of X_1 on $X_2, X_3,...$*. In functional notation this is sometimes written briefly as $X_1 = F(X_2, X_3,...)$ (read "X_1 is a function of X_2, X_3, and so on").

For the case of three variables, the simplest regression equation of X_1 on X_2 and X_3 has the form

$$X_1 = b_{1.23} + b_{12.3} X_2 + b_{13.2} X_3 \qquad (1)$$

where $b_{1.23}$, $b_{12.3}$, and $b_{13.2}$ are constants. If we keep X_3 constant in equation (1), the graph of X_1 versus X_2 is a straight line with slope $b_{12.3}$. If we keep X_2 constant, the graph of X_1 versus X_3 is a straight line with slope $b_{13.2}$. It is clear that the subscripts after the dot indicate the variables held constant in each case.

Due to the fact that X_1 varies partially because of variation in X_2 and partially because of variation in X_3, we call $b_{12.3}$ and $b_{13.2}$ the *partial regression coefficients* of X_1 on X_2 keeping X_3 constant and of X_1 on X_3 keeping X_2 constant, respectively.

Equation (1) is called a *linear regression equation* of X_1 on X_2 and X_3. In a three-dimensional rectangular coordinate system it represents a plane called a *regression plane* and is a generalization of the regression line for two variables, as considered in Chapter 10.

Normal Equations for the Least-Squares Regression Plane

Just as there exist least-squares regression lines approximating a set of N data points (X, Y) in a two-dimensional scatter diagram, so also there exist *least-squares regression planes* fitting a set of N data points (X_1, X_2, X_3) in a three-dimensional scatter diagram.

The least-squares regression plane of X_1 on X_2 and X_3 has the equation (1) where $b_{1.23}$, $b_{12.3}$, and $b_{13.2}$ are determined by solving simultaneously the *normal equations*

$$\sum X_1 = b_{1.23} N + b_{12.3}\sum X_2 + b_{13.2} \sum X_3$$

$$\sum X_1 X_2 = b_{1.23} \sum X_2 + b_{12.3}\sum X_2^2 + b_{13.2} \sum X_2 X_3 \qquad (2)$$

$$\sum X_1 X_3 = b_{1.23} \sum X_3 + b_{12.3}\sum X_2 X_3 + b_{13.2} \sum X_3^2$$

These can be obtained formally by multiplying both sides of equation (1) by 1, X_2, and X_3 successively and summing on both sides.

Unless otherwise specified, whenever we refer to a regression equation it will be assumed that the least-squares regression equation is meant. If $x_1 = X_1 - \bar{X}_1$, $x_2 = \bar{X}_2 - X_2$, and $x_3 = X_3 - \bar{X}_3$, the regression equation of X_1 on X_2 and X_3 can be written more simply as

$$x_1 = b_{12.3}x_2 + b_{13.2}x_3 \tag{3}$$

where $b_{12.3}$ and $b_{13.2}$ are obtained by solving simultaneously the equations

$$\sum x_1 x_2 = b_{12.3} \sum x^2_{} + b_{13.2} \sum x_2 x_3 \tag{4}$$

$$\sum x_1 x_3 = b_{12.3} \sum x_2 x_3 + b_{13.2} \sum x^2_3$$

These equations which are equivalent to the normal equations (2) can be obtained formally by multiplying both sides of equation (3) by x_2 and x_3 successively and summing on both sides.

Regression Planes and Correlation Coefficients

If the linear correlation coefficients between variables X_1 and X_2, X_1 and X_3, and X_2 and X_3, as computed in Chapter 11, are denoted, respectively, by r_{12}, r_{13}, and r_{23} (sometimes called *zero-order correlation coefficients*), then the least-squares regression plane has the equation

$$\frac{x_1}{s_1} = \left(\frac{r_{12} - r_{13}r_{23}}{1 - r^2_{23}} \right) \frac{x_2}{s_2} + \left(\frac{r_{13} - r_{12}r_{23}}{1 - r^2_{23}} \right) \frac{x_3}{s_3} \tag{5}$$

where $x_1 = X - \overline{X}_1$, $x_2 = X_2 - \overline{X}_2$, and $x_3 = X_3 - \overline{X}_3$ and where s_1, s_2, and s_3 are the standard deviations of X_1, X_2, and X_3, respectively.

Standard Error of Estimate

By an obvious generalization of equation (8) of Chapter 11, we can define the *standard error of estimate* of X_1 on X_2 and X_3 by

$$s_{1.23} = \sqrt{\frac{\sum (X_1 - X_{1,\text{est}})^2}{N - 1}} \tag{6}$$

where $X_{1,\text{est}}$ indicates the estimated values of X_1 as calculated from the regression equations (1) or (5).

In terms of the correlation coefficients r_{12}, r_{13}, and r_{23}, the standard error of estimate can also be computed from the result

$$s_{1.23} = s_1 \sqrt{\frac{1 - r_{12}^2 - r_{13}^2 - r_{23}^2 + 2r_{12}r_{13}r_{23}}{1 - r_{23}^2}} \qquad (7)$$

The sampling interpretation of the standard error of estimate for two variables as given in Chapter 11 for the case when N is large can be extended to three dimensions by replacing the lines parallel to the regression line with planes parallel to the regression plane. A better estimate of the population standard error of estimate is given by $\hat{s}_{1.23} = \sqrt{N / (N - 3)} \; s_{123}$.

Coefficient of Multiple Correlation

The coefficient of multiple correlation is defined by an extension of equation (12) or (14) of Chapter 11. In the case of two independent variables, for example, the *coefficient of multiple correlation* is given by

$$R_{1.23} = \sqrt{1 - \frac{s_{1.23}^2}{s_1^2}} \qquad (8)$$

where s_1 is the standard deviation of the variable X_1 and $s_{1.23}$ is given by equation (6) or (7). The quantity $R_{.23}$ is called the *coefficient of multiple determination*.

When a linear regression equation is used, the coefficient of multiple correlation is called the *coefficient of linear multiple correlation*. Unless otherwise specified, whenever we refer to multiple correlation, we shall imply linear multiple correlation.

In terms of r_{12}, r_{13}, and r_{23}, equation (8) can also be written

$$R_{1.23} = \sqrt{\frac{r_{12}^2 + r_{13}^2 - 2r_{12}r_{13}r_{23}}{1 - r_{23}^2}} \qquad (9)$$

A coefficient of multiple correlation, such as $R_{1.23}$, lies between 0 and 1. The closer it is to 1, the better is the linear relationship between the variables. The closer it is to 0, the worse is the linear relationship. If the coefficient of multiple correlation is 1, the correlation is called *perfect*.

Note!

Although a correlation coefficient of 0 indicates no linear relationship between the variables, it is possible that a *nonlinear relationship* may exist.

Change of Dependent Variable

The above results hold when X_1 is considered the dependent variable. However, if we want to consider X_3 (for example) to be the dependent variable instead of X_1, we would only have to replace the subscripts 1 with 3, and 3 with 1, in the formulas already obtained. For example, the regression equation of X_3 on X_1 and X_2 would be

$$\frac{x_3}{s_3} = \left(\frac{r_{23} - r_{13}r_{12}}{1 - r_{12}^2}\right)\frac{x_2}{s_2} + \left(\frac{r_{13} - r_{23}r_{12}}{1 - r_{12}^2}\right)\frac{x_1}{s_1} \qquad (10)$$

as obtained from equation (5), using the results $r_{32} = r_{23}$, $r_{31} = r_{13}$, and $r_{21} = r_{12}$.

Generalizations to More than Three Variables

These are obtained by analogy with the above results. For example, the linear regression equations of X_1 on X_2, X_3, and X_4 can be written

$$X_1 = b_{1.234} + b_{12.34}X_2 + b_{13.24}X_3 + b_{14.23}X_4 \qquad (11)$$

and represents a *hyperplane in four-dimensional space.* By formally multiplying both sides of equation (*11*) by 1, X_2, X_3, and X_4 successively and then summing on both sides, we obtain the normal equations for determining $b_{1.234}$, $b_{12.34}$, $b_{13.24}$, and $b_{14.23}$; substituting these in equation (*11*) then gives us the *least-squares regression equation of X_1 on X_2, X_3, and X_4.* This least-squares regression equation can be written in a form similar to that of equation (5).

Partial Correlation

It is often important to measure the correlation between a dependent variable and one particular independent variable when all other variables involved are kept constant; that is, when the effects of all other variables are removed (often indicated by the phrase "other things being equal"). This can be obtained by defining a *coefficient of partial correlation*, as in equation *(12)* of Chapter 11, except that we must consider the explained and unexplained variations that arise both with and without the particular independent variable.

If we denote by $r_{12.3}$ the coefficient of partial correlation between X_1 and X_2 keeping X_3 constant, we find that

$$r_{12.3} = \frac{r_{12} - r_{13}r_{23}}{\sqrt{(1 - r_{13}^2)(1 - r_{23}^2)}}$$

(12)

Similarly, if $r_{12.3}$ is the coefficient of partial correlation between X_1 and X_2 keeping X_3 and X_4 constant, then

$$r_{12.34} = \frac{r_{12.4} - r_{13.4}r_{23.4}}{\sqrt{(1 - r_{13.4}^2)(1 - r_{23.4}^2)}} = \frac{r_{12.3} - r_{14.3}r_{24.3}}{\sqrt{(1 - r_{14.3}^2)(1 - r_{24.3}^2)}}$$

(13)

These results are useful since by means of them any partial correlation coefficient can ultimately be made to depend on the correlation coefficients r_{12}, r_{23}, etc. (i.e., the *zero-order correlation coefficients*).

In the case of two variables, X and Y, if the two regression lines have equations $Y = a_0 + a_1X$ and $X = b_0 + b_1Y$, we have seen that $r^2 = a_1b_1$. This result can be generalized. For example, if

$$X_1 = b_{1.234} + b_{12.34}X_2 + b_{13.24}X_3 + b_{14.23}X_4$$

(14)

and $\qquad X_4 = b_{4.123} + b_{41.23}X_1 + b_{42.13}X_2 + b_{43.12}X_3$

(15)

are linear regression equations of X_1 on X_2, X_3, and X_4 and of X_4 on X_1, X_2, and X_3, respectively, then

$$r_{4.23} = b_{14.23}b_{41.23}$$

(16)

This can be taken as the starting point for a definition of linear partial correlation coefficients.

Relationships between Multiple and Partial Correlation Coefficients

Interesting results connecting the multiple correlation coefficients can be found. For example, we find that

$$1 - R_{1.23}^2 = (1 - r_{12}^2)(1 - r_{13.2}^2) \qquad (17)$$

$$1 - R_{1.234}^2 = (1 - r_{12}^2)(1 - r_{13.2}^2)(1 - r_{14.23}^2) \qquad (18)$$

Generalizations of these results are easily made.

Chapter 13
ANALYSIS OF VARIANCE

✔ *Two-way Classification, or Two-Factor Experiments*
✔ *Notation for Two-Factor Experiments*
✔ *Variations for Two-Factor Experiments*
✔ *Analysis of Variance for Two-Factor Experiments*
✔ *Two-Factor Experiments with Replication*
✔ *Experimental Design*

The Purpose of Analysis of Variance

In Chapter 5 we used sampling theory to test the significance of differences between two sampling means. We assumed that the two populations from which the samples were drawn had the same variance. In many situations there is a need to test the significance of differences between three or more sampling means or, equivalently, to test the null hypothesis that the sample means are all equal.

EXAMPLE 13.1. Suppose that in an agricultural experiment four different chemical treatments of soil produced mean wheat yields of 28, 22, 18, and 24 bushels per acre, respectively. Is there a significant difference in these means, or is the observed spread due simply to chance?

Problems such as this can be solved by using an important technique known as *analysis of variance*, developed by Fisher. It makes use of the F distribution already considered in Chapter 8.

One-Way Classification or One-Factor Experiments

In a *one-factor experiment*, measurements (or observations) are obtained for a independent groups of samples, where the number of measurements in each group is b. We speak of a *treatments*, each of which has b *repetitions*, or b *replications*. In Example 1, $a = 4$.

The results of a one-factor experiment can be presented in a table having a rows and b columns, as shown in Table 13-1. Here X_{jk} denotes the measurement in the jth row and kth column, where $j = 1, 2,..., a$ and where $k = 1, 2,..., b$. For example, X_{35} refers to the fifth measurement for the third treatment.

Table 13-1

Treatment 1	$X_{11}, X_{12}, ..., X_{1b}$
Treatment 2	$X_{21}, X_{22}, ... , X_{2b}$
$\vdots$	$\vdots$
Treatment a	$X_{a1}, X_{a2} ... , X_{ab}$

We shall denote by $\bar{X}_{j.}$ the mean of the measurements in the jth row. We have

$$\bar{X}_{j.} = \frac{1}{b} \sum_{k=1}^{b} X_{jk} \qquad j = 1, 2, \ldots, a \tag{1}$$

The dot in $\bar{X}_{j.}$ is used to show that the index k has been summed out. The values $\bar{X}_{j.}$ are called *group means*, *treatment means*, or *row means*. The *grand mean*, or *overall mean*, is the mean of all the measurements in all the groups and is denoted by :

$$\bar{X} = \frac{1}{ab} \sum_{j=1}^{a} \sum_{k=1}^{b} X_{jk} \tag{2}$$

Total Variation, Variation within Treatments, and Variation between Treatments

We define the *total variation*, denoted by V, as the sum of the squares of the deviations of each measurement from the grand mean :

$$\text{Total variation} = V = \sum_{j,k} (X_{jk} - \bar{X})^2 \tag{3}$$

By writing the identity

$$X_{jk} - \bar{X} = (X_{jk} - \bar{X}_{j.}) + (\bar{X}_{j.} - \bar{X}) \tag{4}$$

and then squaring and summing over j and k, we have

$$\sum_{j,k} (X_{jk} - \bar{X})^2 = \sum_{j,k} (X_{jk} - \bar{X}_{j.})^2 + \sum_{j,k} (\bar{X}_{j.} - \bar{X})^2 \tag{5}$$

or

$$\sum_{j,k} (X_{jk} - \bar{X})^2 = \sum_{j,k} (X_{jk} - \bar{X}_{j.})^2 + b \sum_{j} (\bar{X}_{j.} - \bar{X})^2 \tag{6}$$

We call the first summation on the right-hand side of equations (5) and (6) the *variation within treatments* (since it involves the squares of the deviations of X_{jk} from the treatment means $X_{j.}$) and denote it by V_W. Thus

$$V_W = \sum_{j,k} (X_{jk} - \bar{X}_{j.})^2 \tag{7}$$

The second summation on the right-hand side of equations (5) and (6) is called the *variation between treatments* (since it involves the squares of the deviations of the various treatment means $X_{j.}$ from the grand mean $\bar{X}$) and is denoted by V_B. Thus

$$V_B = \sum_{j,k} (\bar{X}_{j.} - \bar{X})^2 = b \sum_{j} (\bar{X}_{j.} - \bar{X})^2 \tag{8}$$

Equations (5) and (6) can thus be written

$$V = V_W + V_B \tag{9}$$

Mathematical Model for Analysis of Variance

We can consider each row of Table 13-1 to be a random sample of size b from the population for that particular treatment. The X_{jk} will differ from the population mean μ_j for the jth treatment by a *chance error*, or *random error*, which we denote by ε_{jk}; thus

$$X_{jk} = \mu_j + \varepsilon_{jk} \qquad (10)$$

These errors are assumed to be normally distributed with mean 0 and variance σ^2. If ε is the mean of the population for all treatments and if we let $\alpha_j = \mu_j - \mu$, so that $\mu_j = \mu + \alpha_j$, then equation (10) becomes

$$X_{jk} = \mu + \alpha_j + \varepsilon_{jk} \qquad (11)$$

where $\sum_j \alpha_j = 0$. From equation (11) and the assumption that the ε_{jk} are normally distributed with mean 0 and variance σ^2, we conclude that the X_{jk} can be considered random variables that are normally distributed with mean μ and variance σ^2.

The null hypothesis that all treatment means are equal is given by $(H_0: \mu_j = 0; j = 1, 2, ..., a)$ or, equivalently, by $(H_0: \mu_j = \mu; j = 1, 2, ..., a)$. If H_0 is true, the treatment populations will all have the same normal distribution (i.e., with the same mean and variance). In such case there is just one treatment population (i.e., all treatments are statistically identical); in other words, there is no significant difference between the treatments.

Expected Values of the Variations

It can be shown that the expected values of V_W, V_B, and V are given by

$$E(V_W) = a(b-1)\sigma^2 \qquad (12)$$

$$E(V_B) = (a-1)\sigma^2 + b \sum_j \alpha_j^2 \qquad (13)$$

$$E(V) = (ab-1)\sigma^2 + b \sum_j \alpha_j^2 \qquad (14)$$

From equation (12) it follows that

$$E\left[\frac{V_W}{a(b-1)}\right] = \sigma^2 \qquad (15)$$

so that

$$\hat{S}_W^2 = \frac{V_W}{a(b-1)} \qquad (16)$$

is always a best (unbiased) estimate of σ^2 regardless of whether H_0 is true. On the other hand, we see from equations (12) and (14) that only if H_0 is true (i.e., $\alpha_j = 0$) will we have

$$E\left(\frac{V_B}{a-1}\right) = \sigma^2 \qquad \text{and} \qquad E\left(\frac{V}{ab-1}\right) = \sigma^2 \qquad (17)$$

so that only in such case will

$$\hat{S}_B^2 = \frac{V_B}{a-1} \qquad \text{and} \qquad \hat{S}^2 = \frac{V}{ab-1} \qquad (18)$$

provide unbiased estimates of σ^2. If H_0 is not true, however, then from equation (12) we have

$$E(\hat{S}_B^2) = \sigma^2 + \frac{b}{a-1} \sum_j \alpha_j^2 \qquad (19)$$

The F Test for the Null Hypothesis of Equal Means

If the null hypothesis H_0 is not true (i.e., if the treatment means are not equal), we see from equation (19) that we can expect S_B^2 to be greater than σ^2, with the effect becoming more pronounced as the discrepancy between the means increases. On the other hand, from equations (15) and (16) we can expect S_W^2 to be equal to σ^2 regardless of whether the means are equal. It follows that a good statistic for testing hypothesis H_0 is provided by S_B^2 / S_W^2. If this statistic is significantly large, we can conclude that there is a significant difference between the treatment means and can thus reject H_0; otherwise, we can either accept H_0 or reserve judgment, pending further analysis.

In order to use the S_B^2 / S_W^2 statistic, we must know its sampling distribution. This is provided by Theorem 1.

> **Theorem 1:** The statistic $F = S_B^2 / S_W^2$ has the F distribution with $a - 1$ and $a(b - 1)$ degrees of freedom.

Theorem 1 enables us to test the null hypothesis at some specified significance level by using a one-tailed test of the F distribution (discussed in Chapter 8).

Analysis-of-Variance Tables

The calculations required for the above test are summarized in Table 13-2, which is called an *analysis-of-variance table*. In practice, we would compute V and V_B by using equations (3) and (8) and then by computing $V_W = V - V_B$. It should be noted that the degrees of freedom for the total variation (i.e., $ab - 1$) are equal to the sum of the degrees of freedom for the between-treatments and within-treatments variations.

Table 13-2

Variation	Degrees of Freedom	Mean Square	F
Between treatments, $V_B = b \sum_j (\bar{X}_j - \bar{X})^2$	$a - 1$	$\hat{S}_B^2 = \dfrac{V_B}{a - 1}$	$\dfrac{\hat{S}_B^2}{\hat{S}_W^2}$
Within treatments, $V_W = V - V_B$	$a(b - 1)$	$\hat{S}_W^2 = \dfrac{V_W}{a(b - 1)}$	with $a - 1$ and $a(b - 1)$ degrees of freedom
Total, $V = V_B + V_W$ $= \sum_{j,k} (X_{jk} - \bar{X})^2$	$ab - 1$		

Modifications for Unequal Numbers of Observations

In case the treatments 1,..., a have different numbers of observations — equal to N_1,..., N_a, respectively — the above results are easily modified. Thus we obtain

$$V = \sum_{j,k} (X_{jk} - \bar{X})^2 = \sum_{j,k} X_{jk}^2 - \frac{T^2}{N} \qquad (20)$$

$$V_B = \sum_{j,k} (\bar{X}_{j.} - \bar{X})^2 = \sum_j N_j(\bar{X}_{j.} - \bar{X})^2 = \sum_j \frac{T_j^2}{N_j} - \frac{T^2}{N} \qquad (21)$$

$$V_W = V - V_B \qquad (22)$$

where $\sum_{j,k}$ denotes the summation over k from 1 to N_j and then the summation over j from 1 to a. Table 13-3 is the analysis-of-variance table for this case.

Table 13-3

Variation	Degrees of Freedom	Mean Square	F
Between treatments, $V_B = \sum_j N_j(\bar{X}_{j.} - \bar{X})^2$	$a-1$	$\hat{S}_B^2 = \dfrac{V_B}{a-1}$	$\dfrac{\hat{S}_B^2}{\hat{S}_W^2}$
Within treatments, $V_W = V - V_B$	$N-a$	$\hat{S}_W^2 = \dfrac{V_W}{N-a}$	with $a-1$ and $N-a$ degrees of freedom
Total, $V = V_B + V_W$ $= \sum_{j,k} (X_{jk} - \bar{X})^2$	$N-1$		

Two-Way Classification, or Two-Factor Experiments

The ideas of analysis of variance for one-way classification, or one-factor experiments, can be generalized. Example 13.2 illustrates the procedure for *two-way classification*, or *two-factor experiments*.

EXAMPLE 13.2. Suppose that an agricultural experiment consists of examining the yields per acre of 4 different varieties of wheat, where each variety is grown on 5 different plots of land. Thus a total of (4)(5) = 20 plots are needed. It is convenient in such case to combine the plots into *blocks*, say 4 plots to a block, with a different variety of wheat grown on each plot within a block. Thus 5 blocks would be required here.

In this case there are two classifications, or factors, since there may be differences in yield per acre due to (1) the particular type of wheat grown or (2) the particular block used (which may involve different soil fertility, etc.).

By analogy with the agricultural experiment of Example 13.2, we often refer to the two factors in an experiment as *treatments* and *blocks*, but of course we could simply refer to them as factor 1 and factor 2.

Notation for Two-Factor Experiments

Assuming that we have a treatments and b blocks, we construct Table 13-4, where it is supposed that there is one experimental value (such as yield per acre) corresponding to each treatment and block. For treatment j and block k, we denote this value by X_{jk}. The mean of the entries in the jth row is denoted by $X_{j.}$, where $j = 1,..., a$, while the mean of the entries in the kth column is denoted by $X_{.k}$, where $k = 1,..., b$. The overall, or grand, mean is denoted by X. In symbols,

$$\bar{X}_{j.} = \frac{1}{b}\sum_{k=1}^{b} X_{jk} \qquad \bar{X}_{.k} = \frac{1}{a}\sum_{j=1}^{a} X_{jk} \qquad \bar{X} = \frac{1}{ab}\sum_{j,k} X_{jk} \qquad (23)$$

Table 13-4

	Block 1	2	$\cdots$	b	
Treatment 1	X_{11}	X_{12}	$\cdots$	X_{1b}	$\bar{X}_{1.}$
Treatment 2	X_{21}	X_{22}	$\cdots$	X_{2b}	$\bar{X}_{2.}$
$\vdots$	$\vdots$	$\vdots$	$\vdots$	$\vdots$	$\vdots$
Treatment a	X_{a1}	X_{a2}	$\cdots$	X_{ab}	$\bar{X}_{a.}$
	$\bar{X}_{.1}$	$\bar{X}_{.2}$		$\bar{X}_{.b}$	

Variations for Two-Factor Experiments

As in the case of one-factor experiments, we can define variations for two-factor experiments. We first define the *total variation*, as in equation (*3*), to be

$$V = \sum_{j,k} (X_{jk} - \bar{X})^2 \qquad (24)$$

By writing the identity

$$X_{jk} - \bar{X} = (X_{jk} - X_{j.} - X_{.k} + \bar{X}) + (X_{j.} - X) + (X_{.k} - \bar{X}) \qquad (25)$$

and then squaring and summing over j and k, we can show that

$$V = V_E + V_R + V_C \qquad (26)$$

where

$$V_E = \text{variation due to error or chance} = \sum_{j,k} (X_{jk} - \bar{X}_{j.} - \bar{X}_{.k} + \bar{X})^2$$

$$V_R = \text{variation between rows (treatments)} = b \sum_{j=1}^{a} (\bar{X}_{j.} - \bar{X})^2$$

$$V_C = \text{variation between columns (blocks)} = a \sum_{k=1}^{b} (\bar{X}_{.k} - \bar{X})^2$$

The variation due to error or chance is also known as the *residual variation* or *random variation*.

Analysis of Variance for Two-Factor Experiments

The generalization of the mathematical model for one-factor experiments given by equation (*11*) leads us to assume for two-factor experiments that

$$X_{jk} = \Sigma + \Sigma_j + \Sigma_k + \Sigma_{jk} \qquad (27)$$

where $\sum \alpha_j = 0$ and $\sum \beta_k = 0$. Here μ is the population grand mean, α_j is that part of X_{jk} due to the different treatments (sometimes called the *treatment effects*), β_k is that part of X_{jk} due to the different blocks (sometimes called the *block effects*), and ε_{jk} is that part of X_{jk} due to chance or error. As before, we assume that the ε_{jk} are normally distributed with mean 0 and variance σ^2, so that the X_{jk} are also normally distributed with mean μ and variance σ^2.

Corresponding to results (12), (13), and (14), we can prove that the expectations of the variations are given by

$$E(V_E) = (a-1)(b-1)\sigma^2 \tag{28}$$

$$E(V_R) = (a-1)\sigma^2 + b\sum_j \alpha_j^2 \tag{29}$$

$$E(V_C) = (b-1)\sigma^2 + a\sum_k \beta_k^2 \tag{30}$$

$$E(V) = (ab-1)\sigma^2 + b\sum_j \alpha_j^2 + a\sum_k \beta_k^2 \tag{31}$$

There are two null hypotheses that we would want to test:

$H^{(1)}$: All treatment (row) means are equal; that is, $\alpha_j = 0$, and $j = 1,..., a$.

$H^{(2)}$: All block (column) means are equal; that is, $\beta_k = 0$, and $k = 1,..., b$.

We see from equation (30) that, without regard to $H^{(1)}$ or $H^{(2)}$, a best (unbiased) estimate of σ^2 is provided by

$$\hat{S}_E^2 = \frac{V_E}{(a-1)(b-1)} \quad \text{that is,} \quad E(\hat{S}_E^2) = \sigma^2 \tag{32}$$

Also, if hypotheses $H^{(1)}$ and $H^{(2)}$ are true, then

$$\hat{S}_R^2 = \frac{V_R}{a-1} \qquad \hat{S}_C^2 = \frac{V_C}{b-1} \qquad \hat{S}^2 = \frac{V}{ab-1} \tag{33}$$

will be unbiased estimates of σ^2. If $H^{(1)}$ and $H^{(2)}$ are not true, however, then from equations (28) and (29), respectively, we have

$$E(\hat{S}_R^2) = \sigma^2 + \frac{b}{a-1}\sum_j \alpha_j^2 \tag{34}$$

$$E(\hat{S}_C^2) = \sigma^2 + \frac{a}{b-1}\sum_k \beta_k^2 \tag{35}$$

To test hypothesis $H_0^{(1)}$, it is natural to consider the statistic S_R^2 / S_E^2 since we can see from equation (34) that S_R^2 is expected to differ significantly from σ^2 if the row (treatment) means are significantly different, Similarly, to test hypothesis $H_0^{(2)}$, we consider the statistic S_C^2 / S_E^2. The distributions of S_R^2 / S_E^2 and S_C^2 / S_E^2 are given in Theorem 2, which is analogous to Theorem 1.

> **Theorem 2:** Under hypothesis $H_0^{(1)}$, the statistic S_R^2 / S_E^2 has the F distribution with a - 1 and $(a - 1)(b - 1)$ degrees of freedom. Under hypothesis $H_0^{(1)}$, the statistic S_C^2 / S_E^2 has the F distribution with b - 1 and $(a - 1)$ $(b - 1)$ degrees of freedom.

Theorem 2 enables us to accept or reject $H^{(1)}$ or $H^{(2)}$ at specified significance levels. For convenience, as in the one-factor case, an analysis-of-variance table can be constructed as shown in Table 13-5.

Table 13-5

Variation	Degrees of Freedom	Mean Square	F
Between treatments, $V_R = b \sum_j (\bar{X}_j - \bar{X})^2$	$a - 1$	$\hat{S}_R^2 = \dfrac{V_R}{a - 1}$	$\hat{S}_R^2 / \hat{S}_E^2$ with $a - 1$ and $(a - 1)(b - 1)$ degrees of freedom
Between blocks, $V_C = a \sum_k (\bar{X}_k - \bar{X})^2$	$b - 1$	$\hat{S}_C^2 = \dfrac{V_C}{b - 1}$	$\hat{S}_C^2 / \hat{S}_E^2$ with $b - 1$ and $(a - 1)(b - 1)$ degrees of freedom
Residual or random, $V_E = V - V_R - V_C$	$(a - 1)(b - 1)$	$\hat{S}_E^2 = \dfrac{V_E}{(a - 1)(b - 1)}$	
Total, $V = V_R + V_C + V_E$ $= \sum_{j,k} (X_{jk} - \bar{X})^2$	$ab - 1$		

Two-Factor Experiments with Replication

In Table 13-4 there is only one entry corresponding to a given treatment and a given block. More information regarding the factors can often be obtained by repeating the experiment, a process called *replication*. In such case there will be more than one entry corresponding to a given treatment and a given block. We shall suppose that there are c entries for every position; appropriate changes can be made when the replication numbers are not all equal.

Because of replication, an appropriate model must be used to replace that given by equation (27). We use

$$X_{jkl} = \mu + \alpha_j + \beta_k + \gamma_{jk} + \varepsilon_{jkl} \qquad (36)$$

where the subscripts j, k, and l of X_{jkl} correspond to the jth row (or treatment), the kth column (or block), and the lth repetition (or replication), respectively. In equation (36) the μ, α_j, and β_k are defined as before; ε_{jkl} is a chance or error term, while the γ_{jk} denote the row-column (or treatment-block) *interaction effects*, often simply called *interactions*. We have the restrictions

$$\sum_j \alpha_j = 0 \qquad \sum_k \beta_k = 0 \qquad \sum_j \gamma_{jk} = 0 \qquad \sum_k \gamma_{jk} = 0 \qquad (37)$$

and the X_{jkl} are assumed to be normally distributed with mean μ and variance σ^2.

As before, the total variation V of all the data can be broken up into variations due to rows V_R, columns V_C, interaction V_I, and random or residual error V_E:

$$V = V_R + V_C + V_I + V_E \qquad (38)$$

where

$$V = \sum_{j,k,l} (X_{jkl} - \bar{X})^2 \qquad (39)$$

$$V_R = bc \sum_{j=1}^{a} (\bar{X}_{j..} - \bar{X})^2 \qquad (40)$$

$$V_C = ac \sum_{k=1}^{b} (\bar{X}_{.k.} - \bar{X})^2 \qquad (41)$$

$$V_I = c \sum_{j,k} (\bar{X}_{jk.} - \bar{X}_{j..} - \bar{X}_{.k.} + \bar{X})^2 \tag{42}$$

$$V_E = \sum_{j,k,l} (X_{jkl} - \bar{X}_{jk.})^2 \tag{43}$$

In these results the dots in the subscripts have meanings analogous to those given earlier in the chapter; thus, for example,

$$\bar{X}_{j..} = \frac{1}{bc} \sum_{k,l} X_{jkl} = \frac{1}{b} \sum_{k} \bar{X}_{jk.} \tag{44}$$

The expected values of the variations can be found as before. Using the appropriate number of degrees of freedom for each source of variation, we can set up the analysis-of-variance table as shown in Table 13-6. The F ratios in the last column of Table 13-6 can be used to test the null hypotheses:

$H_0^{(1)}$: All treatment (row) means are equal; that is, $\alpha_j = 0$.
$H_0^{(2)}$: All block (column) means are equal; that is, $\beta_k = 0$.
$H_0^{(3)}$: There are no interactions between treatments and blocks; that is, $\gamma_{jk} = 0$.

Table 13-6

Variation	Degrees of Freedom	Mean Square	F
Between treatments, V_R	$a-1$	$\hat{S}_R^2 = \dfrac{V_R}{a-1}$	$\hat{S}_R^2 / \hat{S}_E^2$ with $a-1$ and $ab(c-1)$ degrees of freedom
Between blocks, V_C	$b-1$	$\hat{S}_C^2 = \dfrac{V_C}{b-1}$	$\hat{S}_C^2 / \hat{S}_E^2$ with $b-1$ and $ab(c-1)$ degrees of freedom
Interaction, V_I	$(a-1)(b-1)$	$\hat{S}_I^2 = \dfrac{V_I}{(a-1)(b-1)}$	$\hat{S}_I^2 / \hat{S}_E^2$ with $(a-1)(b-1)$ and $ab(c-1)$ degrees of freedom
Residual or random, V_E	$ab(c-1)$	$\hat{S}_E^2 = \dfrac{V_E}{ab(c-1)}$	
Total, V	$abc-1$		

From a practical point of view we should first decide whether or not $H_0^{(3)}$ can be rejected at an appropriate level of significance by using the F ratio S_I^2 / S_E^2 of Table 13-6. Two possible cases then arise:

1. $H^{(3)}$ **Cannot Be Rejected.** In this case we can conclude that the interactions are not too large. We can then test $H_0^{(1)}$ and $H_0^{(2)}$ by using the F ratios S_R^2 / S_E^2 and S_B^2 / S_C^2 respectively, as shown in Table 13-6. Some statisticians recommend pooling the variations in this case by taking the total of $V_I + V_E$ and dividing it by the total corresponding degrees of freedom $(a - 1)(b - 1) + ab(c - 1)$ and using this value to replace the denominator S_E^2 in the F test.

2. $H^{(3)}$ **Can Be Rejected.** In this case we can conclude that the interactions are significantly large. Differences in factors would then be of importance only if they were large compared with such interactions. For this reason, many statisticians recommend that $H_0^{(1)}$ and $H_0^{(2)}$ be tested by using the F ratios S_R^2 / S_I^2 and S_C^2 / S_I^2 rather than those given in Table 13-6. We, too, shall use this alternative procedure.

The analysis of variance with replication is most easily performed by first totaling the replication values that correspond to particular treatments (rows) and blocks (columns). This produces a two factor table with single entries, which can be analyzed as in Table 13-5.

Experimental Design

The techniques of analysis of variance discussed above are employed after the results of an experiment have been obtained. However, in order to gain as much information as possible, the design of an experiment must be planned carefully in advance; this is often referred to as the *design of the experiment*. The following are some important examples of experimental design:

1. **Complete Randomization.** Suppose that we have an agricultural experiment as in Example 13.1. To design such an experiment, we could divide the land into $4 \times 4 = 16$ plots (indicated in Fig. 13-1 by squares, although physically any shape can be used) and assign each treatment (indicated by A, B, C, and D) to four blocks chosen completely at random. The purpose of the randomization is to eliminate various sources of error, such as soil fertility.

D	A	C	C
B	D	B	A
D	C	B	D
A	B	C	A

Complete
Randomization
Figure 13-1

I	C	B	A	D
II	A	B	D	C
III	B	C	D	A
IV	A	D	C	B

Randomized
Blocks
Figure 13-2

D	B	C	A
B	D	A	C
C	A	D	B
A	C	B	D

Latin
Square

Figure 13-3

B_γ	A_β	D_δ	C_α
A_δ	B_α	C_γ	D_β
D_α	C_δ	B_β	A_γ
C_β	D_γ	A_α	B_δ

Graeco-Latin
Square

Figure 13-4

2. **Randomized Blocks.** When, as in Example 13.2, it is necessary to have a complete set of treatments for each block, the treatments A, B, C, and D are introduced in random order within each block: I, II, III, and IV (i.e., the rows in Fig. 13-2), and for this reason the blocks are referred to as *randomized blocks*. This type of design is used when it is desired to control *one source of error or variability*: namely, the difference in blocks.

3. **Latin Squares.** For some purposes it is necessary to control *two sources of error or variability* at the same time, such as the difference in rows and the difference in columns. In the experiment of Example 13.1, for instance, errors in different rows and columns could be due to changes in soil fertility in different parts of the land. In such case it is desirable that each treatment occur once in each row and once in each column, as in Fig. 13-3. The arrangement is called a *Latin square* from the fact that the Latin letters A, B, C, and D are used.

4. **Graeco-Latin Squares.** If it is necessary to control *three sources of error or variability*, a *Graeco-Latin square* is used, as shown in Fig. 13-4. Such a square is essentially two Latin squares superimposed on each other, with the Latin letters A, B, C, and D used for one square and the Greek letters α, β, γ, and δ used for the other square. The additional requirement that must be met is that each Latin letter must be used once and only once with each Greek letter; when this requirement is met, the square is said to be *orthogonal*.

Chapter 14
NONPARAMETRIC TESTS

IN THIS CHAPTER:

Introduction

Most tests of hypotheses and significance (or decision rules) considered in previous chapters require various assumptions about the distribution of the population from which the samples are drawn.

Situations arise in practice in which such assumptions may not be justified or in which there is doubt that they apply, as in the case where a population may be highly skewed. Because of this, statisticians have devised various tests and methods that are independent of

population distributions and associated parameters. These are called *nonparametric tests*.

Nonparametric tests can be used as shortcut replacements for more complicated tests. They are especially valuable in dealing with nonnumerical data, such as arise when consumers rank cereals or other products in order of preference.

The Sign Test

Consider Table 14-1, which shows the numbers of defective bolts produced by two different types of machines (I and II) on 12 consecutive days and which assumes that the machines have the same total output per day. We wish to test the hypothesis H_0 that there is no difference between the machines: that the observed differences between the machines in terms of the numbers of defective bolts they produce are merely the result of chance, which is to say that the samples come from the same population. A simple nonparametric test in the case of such paired samples is provided by the *sign test*. This test consists of taking the difference between the numbers of defective bolts for each day and writing only the *sign* of the difference; for instance, for day 1 we have 47 - 71, which is negative. In this way we obtain from Table 14-1 the sequence of signs

$$- - + - - - + - + - - - \qquad (1)$$

(i.e., 3 pluses and 9 minuses). Now if it is just as likely to get a + as a -, we would expect to get 6 of each. The test of H_0 is thus equivalent to that of whether a coin is fair if 12 tosses result in 3 heads (+) and 9 tails (-). This involves the binomial distribution of Chapter 4.

Table 14-1

Day	1	2	3	4	5	6	7	8	9	10	11	12
Machine I	47	56	54	49	36	48	51	38	61	49	56	52
Machine II	71	63	45	64	50	55	42	46	53	57	75	60

Remark 1: If on some day the machines produced the same number of defective bolts, a difference of *zero* would appear in sequence (*1*). In such case we can omit these sample values and use 11 instead of 12 observations.

Remark 2: A normal approximation to the binomial distribution, using a correction for continuity, can also be used.

Although the sign test is particularly useful for paired samples, as in Table 14-1, it can also be used for problems involving single samples.

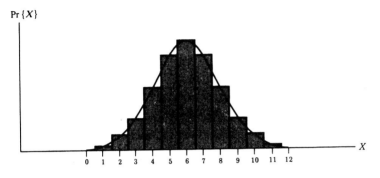

Figure 14-1

EXAMPLE 14.1. Referring to Table 14-1, to test the hypothesis H_0 that there is no difference between machines I and II against the alternative hypothesis H_1 that there is a difference at the 0.05 significance level we turn to Figure 14-1, which is a graph of the binomial distribution (and a normal approximation to it) that gives the probabilities of X heads in 12 tosses of a fair coin, where $X = 0, 1, 2,..., 12$. From Chapter 3 the probability of X heads is

$$\Pr\{X\} = \binom{12}{X}\left(\frac{1}{2}\right)^X \left(\frac{1}{2}\right)^{12-X} = \binom{12}{X}\left(\frac{1}{2}\right)^{12}$$

whereby $\Pr\{0\} = 0.00024$, $\Pr\{1\} = 0.00293$, $\Pr\{2\} = 0.01611$, and $\Pr\{3\} = 0.05371$.

Since H_1 is the hypothesis that there is a *difference* between the machines, rather than the hypothesis that machine I is *better* than machine II, we use a two-tailed test. For the 0.05 significance level, each tail has the associated probability (0.05) = 0.025. We now add the probabilities in the left-hand tail until the sum exceeds 0.025. Thus

$$\Pr\{0, 1, \text{ or } 2 \text{ heads}\} = 0.00024 + 0.00293 + 0.01611 = 0.01928$$

$$\begin{aligned}\Pr\{0, 1, 2, \text{ or } 3 \text{ heads}\} &= 0.00024 + 0.00293 + 0.01611 + 0.05371 \\ &= 0.07299\end{aligned}$$

Since 0.025 is greater than 0.01928 but less than 0.07299, we can reject hypothesis H_0 if the number of heads is 2 or less (or, by symmetry, if the number of heads is 10 or more); however, the number of heads [the + signs in sequence (1)] is 3. Thus we cannot reject H_0 at the 0.05 level and must conclude that there is no difference between the machines at this level.

The Mann-Whitney U Test

Consider Table 14-2, which shows the strengths of cables made from two different alloys, I and II. In this table we have two samples: 8 cables of alloy I and 10 cables of alloy II. We would like to decide whether or not there is a difference between the samples or, equivalently, whether or not they come from the same population. Although this problem can be worked by using the *t* test of Chapter 8, a nonparametric test called the *Mann-Whitney U test*, or briefly the *U test*, is useful. This test consists of the following steps:

Table 14-2

Alloy I				Alloy II				
18.3	16.4	22.7	17.8	12.6	14.1	20.5	10.7	15.9
18.9	25.3	16.1	24.2	19.6	12.9	15.2	11.8	14.7

Step 1. Combine all sample values in an array from the smallest to the largest, and assign ranks (in this case from 1 to 18) to all these values. If two or more sample values are identical (i.e., there are *tie scores*, or briefly *ties*), the sample values are each assigned a rank equal to the

mean of the ranks that would otherwise be assigned. If the entry 18.9 in Table 14-2 were 18.3, two identical values 18.3 would occupy ranks 12 and 13 in the array so that the rank assigned to each would be (12+ 13) = 12.5.

Step 2. Find the sum of the ranks for each of the samples, Denote these sums by R_1 and R_2, where N_1 and N_2 are the respective sample sizes. For convenience, choose N_1 as the smaller size if they are unequal, so that $N_1 \leq N_2$. A significant difference between the rank sums R_1 and R_2 implies a significant difference between the samples.

Step 3. To test the difference between the rank sums, use the statistic

$$U = N_1 N_2 + \frac{N_1(N_1+1)}{2} - R_1 \qquad (2)$$

corresponding to sample 1. The sampling distribution of U is symmetrical and has a mean and variance given, respectively, by the formulas

$$\mu_U = \frac{N_1(N_1 + N_2 + 1)}{2} \qquad \sigma^2_U = \frac{N_1 N_2(N_1 + N_2 + 1)}{12} \qquad (3)$$

If N_1 and N_2 are both at least equal to 8, it turns out that the distribution of U is nearly normal, so that

$$z = \frac{U - \mu_U}{\sigma_U} \qquad (4)$$

is normally distributed with mean 0 and variance 1. Using the z table in Appendix A, we can then decide whether the samples are significantly different.

Remark 3: A value corresponding to sample 2 is given by the statistic

$$U = N_1 N_2 + \frac{N_2(N_2+1)}{2} - R_2 \qquad (5)$$

and has the same sampling distribution as statistic (2), with the mean and variance of formulas (3). Statistic (5) is related to statistic (2), for if U_1 and U_2 are the values corresponding to statistics (2) and (5), respectively, then we have the result

$$U_1 + U_2 = N_1 N_2 \tag{6}$$

We also have

$$R_1 + R_2 = \frac{N(N+1)}{2} \tag{7}$$

where $N = N_1 + N_2$. Result (7) can provide a check for calculations.

Remark 4: The statistic U in equation (2) is the total number of times that sample 1 values precede sample 2 values when all sample values are arranged in increasing order of magnitude. This provides an alternative *counting method* for finding U.

The Kruskal-Wallis *H* Test

The U test is a nonparametric test for deciding whether or not two samples come from the same population. A generalization of this for k samples is provided by the *Kruskal-Wallis H* test, or briefly the *H* test.

This test may be described thus: Suppose that we have k samples of sizes N_1, N_2,..., N_K, with the total size of all samples taken together being given by $N = N_1 + N_2 + \dots + N_k$. Suppose further that the data from all the samples taken together are ranked and that the sums of the ranks for the k samples are R_1, R_2..., R_k, respectively. If we define the statistic

$$H = \frac{12}{N(N+1)} \sum_{j=1}^{k} \frac{R_j^2}{N_j} - 3(N+1) \tag{8}$$

then it can be shown that the sampling distribution of H is very nearly a *chi-square distribution* with k - 1 degrees of freedom, provided that N_1, N_2,..., N_k are all at least 5.

Note!

The *H* test provides a nonparametric method in the *analysis of variance* for one-way classification, or one-factor experiments, and generalizations can be made.

The *H* Test Corrected for Ties

In case there are too many ties among the observations in the sample data, the value of *H* given by statistic (*8*) is smaller than it should be. The corrected value of *H*, denoted by H_c, is obtained by dividing the value given in statistic (*8*) by the correction factor

$$1 - \frac{\sum (T^3 - T)}{N^3 - N} \qquad (9)$$

where *T* is the number of ties corresponding to each observation and where the sum is taken over all the observations. If there are no ties, then *T* = 0 and factor (*9*) reduces to 1, so that no correction is needed. In practice, the correction is usually negligible (i.e., it is not enough to warrant a change in the decision).

Spearman's Rank Correlation

Nonparametric methods can also be used to measure the correlation of two variables, *X* and *Y*. Instead of using precise values of the variables, or when such precision is unavailable, the data may be ranked from 1 to *N* in order of size, importance, etc. If *X* and *Y* are ranked in such a manner, the *coefficient of rank correlation*, or *Spearman's formula for rank correlation* (as it is often called), is given by

$$r_S = 1 - \frac{6 \sum D^2}{N(N^2 - 1)} \qquad (10)$$

where *D* denotes the differences between the ranks of corresponding values of *X* and *Y*, and where *N* is the number of pairs of values (*X*, *Y*) in the data assuming there are no ties.

Appendix A

Areas under the Standard Normal Curve from 0 to z

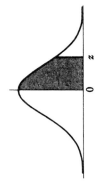

z	0	1	2	3	4	5	6	7	8	9
0.0	.0000	.0040	.0080	.0120	.0160	.0199	.0239	.0279	.0319	.0359
0.1	.0398	.0438	.0478	.0517	.0557	.0596	.0636	.0675	.0714	.0754
0.2	.0793	.0832	.0871	.0910	.0948	.0987	.1026	.1064	.1103	.1141
0.3	.1179	.1217	.1255	.1293	.1331	.1368	.1406	.1443	.1480	.1517
0.4	.1554	.1591	.1628	.1664	.1700	.1736	.1772	.1808	.1844	.1879
0.5	.1915	.1950	.1985	.2019	.2054	.2088	.2123	.2157	.2190	.2224
0.6	.2258	.2291	.2324	.2357	.2389	.2422	.2454	.2486	.2518	.2549
0.7	.2580	.2612	.2642	.2673	.2704	.2734	.2764	.2794	.2823	.2852
0.8	.2881	.2910	.2939	.2967	.2996	.3023	.3051	.3078	.3106	.3133
0.9	.3159	.3186	.3212	.3238	.3264	.3289	.3315	.3340	.3365	.3389
1.0	.3413	.3438	.3461	.3485	.3508	.3531	.3554	.3577	.3599	.3621
1.1	.3643	.3665	.3686	.3708	.3729	.3749	.3770	.3790	.3810	.3830
1.2	.3849	.3869	.3888	.3907	.3925	.3944	.3962	.3980	.3997	.4015
1.3	.4032	.4049	.4066	.4082	.4099	.4115	.4131	.4147	.4162	.4177
1.4	.4192	.4207	.4222	.4236	.4251	.4265	.4279	.4292	.4306	.4319

128

z										
1.5	.4332	.4345	.4357	.4370	.4382	.4394	.4406	.4418	.4429	.4441
1.6	.4452	.4463	.4474	.4484	.4495	.4505	.4515	.4525	.4535	.4545
1.7	.4554	.4564	.4573	.4582	.4591	.4599	.4608	.4616	.4625	.4633
1.8	.4641	.4649	.4656	.4664	.4671	.4678	.4686	.4693	.4699	.4706
1.9	.4713	.4719	.4726	.4732	.4738	.4744	.4750	.4756	.4761	.4767
2.0	.4772	.4778	.4783	.4788	.4793	.4798	.4803	.4808	.4812	.4817
2.1	.4821	.4826	.4830	.4834	.4838	.4842	.4846	.4850	.4854	.4857
2.2	.4861	.4864	.4868	.4871	.4875	.4878	.4881	.4884	.4887	.4890
2.3	.4893	.4896	.4898	.4901	.4904	.4906	.4909	.4911	.4913	.4916
2.4	.4918	.4920	.4922	.4925	.4927	.4929	.4931	.4932	.4934	.4936
2.5	.4938	.4940	.4941	.4943	.4945	.4946	.4948	.4949	.4951	.4952
2.6	.4953	.4955	.4956	.4957	.4959	.4960	.4961	.4962	.4963	.4964
2.7	.4965	.4966	.4967	.4968	.4969	.4970	.4971	.4972	.4973	.4974
2.8	.4974	.4975	.4976	.4977	.4977	.4978	.4979	.4979	.4980	.4981
2.9	.4981	.4982	.4982	.4983	.4984	.4984	.4985	.4985	.4986	.4986
3.0	.4987	.4987	.4987	.4988	.4988	.4989	.4989	.4989	.4990	.4990
3.1	.4990	.4991	.4991	.4991	.4992	.4992	.4992	.4992	.4993	.4993
3.2	.4993	.4993	.4994	.4994	.4994	.4994	.4994	.4995	.4995	.4995
3.3	.4995	.4995	.4995	.4996	.4996	.4996	.4996	.4996	.4996	.4997
3.4	.4997	.4997	.4997	.4997	.4997	.4997	.4997	.4997	.4997	.4998
3.5	.4998	.4998	.4998	.4998	.4998	.4998	.4998	.4998	.4998	.4998
3.6	.4998	.4998	.4999	.4999	.4999	.4999	.4999	.4999	.4999	.4999
3.7	.4999	.4999	.4999	.4999	.4999	.4999	.4999	.4999	.4999	.4999
3.8	.4999	.4999	.4999	.4999	.4999	.4999	.4999	.4999	.4999	.4999
3.9	.5000	.5000	.5000	.5000	.5000	.5000	.5000	.5000	.5000	.5000

Appendix B

Percentile Values (t_p) for Student's t Distribution
with υ Degrees of Freedom
(shaded area = p)

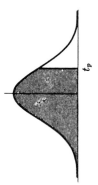

υ	$t_{.995}$	$t_{.99}$	$t_{.975}$	$t_{.95}$	$t_{.90}$	$t_{.80}$	$t_{.75}$	$t_{.70}$	$t_{.60}$	$t_{.55}$
1	63.66	31.82	12.71	6.31	3.08	1.376	1.000	.727	.325	.158
2	9.92	6.96	4.30	2.92	1.89	1.061	.816	.617	.289	.142
3	5.84	4.54	3.18	2.35	1.64	.978	.765	.584	.277	.137
4	4.60	3.75	2.78	2.13	1.53	.941	.741	.569	.271	.134
5	4.03	3.36	2.57	2.02	1.48	.920	.727	.559	.267	.132
6	3.71	3.14	2.45	1.94	1.44	.906	.718	.553	.265	.131
7	3.50	3.00	2.36	1.90	1.42	.896	.711	.549	.263	.130
8	3.36	2.90	2.31	1.86	1.40	.889	.706	.546	.262	.130
9	3.25	2.82	2.26	1.83	1.38	.883	.703	.543	.261	.129
10	3.17	2.76	2.23	1.81	1.37	.879	.700	.542	.260	.129
11	3.11	2.72	2.20	1.80	1.36	.876	.697	.540	.260	.129
12	3.06	2.68	2.18	1.78	1.36	.873	.695	.539	.259	.128
13	3.01	2.65	2.16	1.77	1.35	.870	.694	.538	.259	.128
14	2.98	2.62	2.14	1.76	1.34	.868	.692	.537	.258	.128

15	2.95	2.60	2.13	1.75	1.34	.866	.691	.536	.258	.128
16	2.92	2.58	2.12	1.75	1.34	.865	.690	.535	.258	.128
17	2.90	2.57	2.11	1.74	1.33	.863	.689	.534	.257	.128
18	2.88	2.55	2.10	1.73	1.33	.862	.688	.534	.257	.127
19	2.86	2.54	2.09	1.73	1.33	.861	.688	.533	.257	.127
20	2.84	2.53	2.09	1.72	1.32	.860	.687	.533	.257	.127
21	2.83	2.52	2.08	1.72	1.32	.859	.686	.532	.257	.127
22	2.82	2.51	2.07	1.72	1.32	.858	.686	.532	.256	.127
23	2.81	2.50	2.07	1.71	1.32	.858	.685	.532	.256	.127
24	2.80	2.49	2.06	1.71	1.32	.857	.685	.531	.256	.127
25	2.79	2.48	2.06	1.71	1.32	.856	.684	.531	.256	.127
26	2.78	2.48	2.06	1.71	1.32	.856	.684	.531	.256	.127
27	2.77	2.47	2.05	1.70	1.31	.855	.684	.531	.256	.127
28	2.76	2.47	2.05	1.70	1.31	.855	.683	.530	.256	.127
29	2.76	2.46	2.04	1.70	1.31	.854	.683	.530	.256	.127
30	2.75	2.46	2.04	1.70	1.31	.854	.683	.530	.256	.127
40	2.70	2.42	2.02	1.68	1.30	.851	.681	.529	.255	.126
60	2.66	2.39	2.00	1.67	1.30	.848	.679	.527	.254	.126
120	2.62	2.36	1.98	1.66	1.29	.845	.677	.526	.254	.126
∞	2.58	2.33	1.96	1.645	1.28	.842	.674	.524	.253	.126

Source: R. A. Fisher and F. Yates, *Statistical Tables for Biological, Agricultural and Medical Research* (5th edition), Table III, Oliver and Boyd Ltd., Edinburgh, by permission of the authors and publishers.

Appendix C

Percentile Values (χ_p^2) for the Chi-Square Distribution

with υ_1 Degrees of Freedom

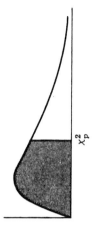

υ	$\chi^2_{.995}$	$\chi^2_{.99}$	$\chi^2_{.975}$	$\chi^2_{.95}$	$\chi^2_{.90}$	$\chi^2_{.75}$	$\chi^2_{.50}$	$\chi^2_{.25}$	$\chi^2_{.10}$	$\chi^2_{.05}$	$\chi^2_{.025}$	$\chi^2_{.01}$	$\chi^2_{.005}$
1	7.88	6.63	5.02	3.84	2.71	1.32	.455	.102	.0158	.0039	.0010	.0002	.0000
2	10.6	9.21	7.38	5.99	4.61	2.77	1.39	.575	.211	.103	.0506	.0201	.0100
3	12.8	11.3	9.35	7.81	6.25	4.11	2.37	1.21	.584	.352	.216	.115	.072
4	14.9	13.3	11.1	9.49	7.78	5.39	3.36	1.92	1.06	.711	.484	.297	.207
5	16.7	15.1	12.8	11.1	9.24	6.63	4.35	2.67	1.61	1.15	.831	.554	.412
6	18.5	16.8	14.4	12.6	10.6	7.84	5.35	3.45	2.20	1.64	1.24	.872	.676
7	20.3	18.5	16.0	14.1	12.0	9.04	6.35	4.25	2.83	2.17	1.69	1.24	.989
8	22.0	20.1	17.5	15.5	13.4	10.2	7.34	5.07	3.49	2.73	2.18	1.65	1.34
9	23.6	21.7	19.0	16.9	14.7	11.4	8.34	5.90	4.17	3.33	2.70	2.09	1.73
10	25.2	23.2	20.5	18.3	16.0	12.5	9.34	6.74	4.87	3.94	3.25	2.56	2.16
11	26.8	24.7	21.9	19.7	17.3	13.7	10.3	7.58	5.58	4.57	3.82	3.05	2.60
12	28.3	26.2	23.3	21.0	18.5	14.8	11.3	8.44	6.30	5.23	4.40	3.57	3.07
13	29.8	27.7	24.7	22.4	19.8	16.0	12.3	9.30	7.04	5.89	5.01	4.11	3.57
14	31.3	29.1	26.1	23.7	21.1	17.1	13.3	10.2	7.79	6.57	5.63	4.66	4.07

df													
15	4.60	5.23	6.26	7.26	8.55	11.0	14.3	18.2	22.3	25.0	27.5	30.6	32.8
16	5.14	5.81	6.91	7.96	9.31	11.9	15.3	19.4	23.5	26.3	28.8	32.0	34.3
17	5.70	6.41	7.56	8.67	10.1	12.8	16.3	20.5	24.8	27.6	30.2	33.4	35.7
18	6.26	7.01	8.23	9.39	10.9	13.7	17.3	21.6	26.0	28.9	31.5	34.8	37.2
19	6.84	7.63	8.91	10.1	11.7	14.6	18.3	22.7	27.2	30.1	32.9	36.2	38.6
20	7.43	8.26	9.59	10.9	12.4	15.5	19.3	23.8	28.4	31.4	34.2	37.6	40.0
21	8.03	8.90	10.3	11.6	13.2	16.3	20.3	24.9	29.6	32.7	35.5	38.9	41.4
22	8.64	9.54	11.0	12.3	14.0	17.2	21.3	26.0	30.8	33.9	36.8	40.3	42.8
23	9.26	10.2	11.7	13.1	14.8	18.1	22.3	27.1	32.0	35.2	38.1	41.6	44.2
24	9.89	10.9	12.4	13.8	15.7	19.0	23.3	28.2	33.2	36.4	39.4	43.0	45.6
25	10.5	11.5	13.1	14.6	16.5	19.9	24.3	29.3	34.4	37.7	40.6	44.3	46.9
26	11.2	12.2	13.8	15.4	17.3	20.8	25.3	30.4	35.6	38.9	41.9	45.6	48.3
27	11.8	12.9	14.6	16.2	18.1	21.7	26.3	31.5	36.7	40.1	43.2	47.0	49.6
28	12.5	13.6	15.3	16.9	18.9	22.7	27.3	32.6	37.9	41.3	44.5	48.3	51.0
29	13.1	14.3	16.0	17.7	19.8	23.6	28.3	33.7	39.1	42.6	45.7	49.6	52.3
30	13.8	15.0	16.8	18.5	20.6	24.5	29.3	34.8	40.3	43.8	47.0	50.9	53.7
40	20.7	22.2	24.4	26.5	29.1	33.7	39.3	45.6	51.8	55.8	59.3	63.7	66.8
50	28.0	29.7	32.4	34.8	37.7	42.9	49.3	56.3	63.2	67.5	71.4	76.2	79.5
60	35.5	37.5	40.5	43.2	46.5	52.3	59.3	67.0	74.4	79.1	83.3	88.4	92.0
70	43.3	45.4	48.8	51.7	55.3	61.7	69.3	77.6	85.5	90.5	95.0	100.4	104.2
80	51.2	53.5	57.2	60.4	64.3	71.1	79.3	88.1	96.6	101.9	106.6	112.3	116.3
90	59.2	61.8	65.6	69.1	73.3	80.6	89.3	98.6	107.6	113.1	118.1	124.1	128.3
100	67.3	70.1	74.2	77.9	82.4	90.1	99.3	109.1	118.5	124.3	129.6	135.8	140.2

Source: Catherine M. Thompson, Table of percentage points of the χ^2 distribution, Biometrika, Vol. 32 (1941), by permission of the author and publisher.

Appendix D

99th Percentile Values for the F Distribution

(υ_1 degrees of freedom in numerator)
(υ_2 degrees of freedom in Denominator)

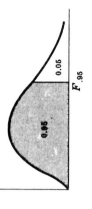

υ_1 / υ_2	1	2	3	4	5	6	7	8	9	10	12	15	20	24	30	40	60	120	∞
1	161	200	216	225	230	234	237	239	241	242	244	246	248	249	250	251	252	253	254
2	18.5	19.0	19.2	19.2	19.3	19.3	19.4	19.4	19.4	19.4	19.4	19.4	19.4	19.5	19.5	19.5	19.5	19.5	19.5
3	10.1	9.55	9.28	9.12	9.01	8.94	8.89	8.85	8.81	8.79	8.74	8.70	8.66	8.64	8.62	8.59	8.57	8.55	8.53
4	7.71	6.94	6.59	6.39	6.26	6.16	6.09	6.04	6.00	5.96	5.91	5.86	5.80	5.77	5.75	5.72	5.69	5.66	5.63
5	6.61	5.79	5.41	5.19	5.05	4.95	4.88	4.82	4.77	4.74	4.68	4.62	4.56	4.53	4.50	4.46	4.43	4.40	4.37
6	5.99	5.14	4.76	4.53	4.39	4.28	4.21	4.15	4.10	4.06	4.00	3.94	3.87	3.84	3.81	3.77	3.74	3.70	3.67
7	5.59	4.74	4.35	4.12	3.97	3.87	3.79	3.73	3.68	3.64	3.57	3.51	3.44	3.41	3.38	3.34	3.30	3.27	3.23
8	5.32	4.46	4.07	3.84	3.69	3.58	3.50	3.44	3.39	3.35	3.28	3.22	3.15	3.12	3.08	3.04	3.01	2.97	2.93
9	5.12	4.26	3.86	3.63	3.48	3.37	3.29	3.23	3.18	3.14	3.07	3.01	2.94	2.90	2.86	2.83	2.79	2.75	2.71
10	4.96	4.10	3.71	3.48	3.33	3.22	3.14	3.07	3.02	2.98	2.91	2.85	2.77	2.74	2.70	2.66	2.62	2.58	2.54
11	4.84	3.98	3.59	3.36	3.20	3.09	3.01	2.95	2.90	2.85	2.79	2.72	2.65	2.61	2.57	2.53	2.49	2.45	2.40
12	4.75	3.89	3.49	3.26	3.11	3.00	2.91	2.85	2.80	2.75	2.69	2.62	2.54	2.51	2.47	2.43	2.38	2.34	2.30
13	4.67	3.81	3.41	3.18	3.03	2.92	2.83	2.77	2.71	2.67	2.60	2.53	2.46	2.42	2.38	2.34	2.30	2.25	2.21

$F_{.95}$

df																			
14	4.60	3.74	3.34	3.11	2.96	2.85	2.76	2.70	2.65	2.60	2.53	2.46	2.39	2.35	2.31	2.27	2.22	2.18	2.13
15	4.54	3.68	3.29	3.06	2.90	2.79	2.71	2.64	2.59	2.54	2.48	2.40	2.33	2.29	2.25	2.20	2.16	2.11	2.07
16	4.49	3.63	3.24	3.01	2.85	2.74	2.66	2.59	2.54	2.49	2.42	2.35	2.28	2.24	2.19	2.15	2.11	2.06	2.01
17	4.45	3.59	3.20	2.96	2.81	2.70	2.61	2.55	2.49	2.45	2.38	2.31	2.23	2.19	2.15	2.10	2.06	2.01	1.96
18	4.41	3.55	3.16	2.93	2.77	2.66	2.58	2.51	2.46	2.41	2.34	2.27	2.19	2.15	2.11	2.06	2.02	1.97	1.92
19	4.38	3.52	3.13	2.90	2.74	2.63	2.54	2.48	2.42	2.38	2.31	2.23	2.16	2.11	2.07	2.03	1.98	1.93	1.88
20	4.35	3.49	3.10	2.87	2.71	2.60	2.51	2.45	2.39	2.35	2.28	2.20	2.12	2.08	2.04	1.99	1.95	1.90	1.84
21	4.32	3.47	3.07	2.84	2.68	2.57	2.49	2.42	2.37	2.32	2.25	2.18	2.10	2.05	2.01	1.96	1.92	1.87	1.81
22	4.30	3.44	3.05	2.82	2.66	2.55	2.46	2.40	2.34	2.30	2.23	2.15	2.07	2.03	1.98	1.94	1.89	1.84	1.78
23	4.28	3.42	3.03	2.80	2.64	2.53	2.44	2.37	2.32	2.27	2.20	2.13	2.05	2.01	1.96	1.91	1.86	1.81	1.76
24	4.26	3.40	3.01	2.78	2.62	2.51	2.42	2.36	2.30	2.25	2.18	2.11	2.03	1.98	1.94	1.89	1.84	1.79	1.73
25	4.24	3.39	2.99	2.76	2.60	2.49	2.40	2.34	2.28	2.24	2.16	2.09	2.01	1.96	1.92	1.87	1.82	1.77	1.71
26	4.23	3.37	2.98	2.74	2.59	2.47	2.39	2.32	2.27	2.22	2.15	2.07	1.99	1.95	1.90	1.85	1.80	1.75	1.69
27	4.21	3.35	2.96	2.73	2.57	2.46	2.37	2.31	2.25	2.20	2.13	2.06	1.97	1.93	1.88	1.84	1.79	1.73	1.67
28	4.20	3.34	2.95	2.71	2.56	2.45	2.36	2.29	2.24	2.19	2.12	2.04	1.96	1.91	1.87	1.82	1.77	1.71	1.65
29	4.18	3.33	2.93	2.70	2.55	2.43	2.35	2.28	2.22	2.18	2.10	2.03	1.94	1.90	1.85	1.81	1.75	1.70	1.64
30	4.17	3.32	2.92	2.69	2.53	2.42	2.33	2.27	2.21	2.16	2.09	2.01	1.93	1.89	1.84	1.79	1.74	1.68	1.62
40	4.08	3.23	2.84	2.61	2.45	2.34	2.25	2.18	2.12	2.08	2.00	1.92	1.84	1.79	1.74	1.69	1.64	1.58	1.51
60	4.00	3.15	2.76	2.53	2.37	2.25	2.17	2.10	2.04	1.99	1.92	1.84	1.75	1.70	1.65	1.59	1.53	1.47	1.39
120	3.92	3.07	2.68	2.45	2.29	2.18	2.09	2.02	1.96	1.91	1.83	1.75	1.66	1.61	1.55	1.50	1.43	1.35	1.25
∞	3.84	3.00	2.60	2.37	2.21	2.10	2.01	1.94	1.88	1.83	1.75	1.67	1.57	1.52	1.46	1.39	1.32	1.22	1.00

Source: E. S. Pearson and H. O. Hartley, *Biometrika Tables for Statisticians*, Vol. 2 (1972), Table 5, page 178, by permission.

Index

Probable error, 51
Product moment formula, 93–94
Proportions, 59–60

Quartiles, 16

Random numbers, 39–40
Random samples, 39–40
Raw data, 6
Rectangular coordinates, 5–6
Regression, 83, 86, 88–89, 94–95, 97–99
Relative dispersion, 19
Relative-frequency, 9–10, 21–22
Replacement, 40

Sample mean, 28
Sample tests of difference, 59–60
Sampling
    correlation, 91
    difference and sums, 42–44
    elementary theory, 38–39
    proportions, 42
    random, 39–40
    regression, 94–95
    replacement, 40
    small, 61–68
Scientific notation, 3–4
Sign test, 12–24
Significance, 53–54, 55, 64–65
Small samples, 61–62
Spearman's rank correlation, 127
Special test, 58
Standard deviation, 17, 18, 51
Standard error of estimate, 90–91, 99–100
Standard errors, 44
Standard score, 19

Standardized variable, 19
Straight line, 79
Student's *t* distribution, 62–63
Tests
    goodness of fit, 72
    hypotheses, 53–54, 64–65
    Kruskal-Wallis H, 126–27
    Mann-Whitney U, 124–26
    non-parametric, 121–27
    normal distribution, 55–57
    one-tailed, 57
    sample differences, 59–60
    significance, 64–65, 71–72
    sign, 122–24
    special, 58
    two-tailed, 57
Two-tailed tests, 57
Two-way classification, 111–18
Two-way experiments, 111–18
Type I errors, 54
Type II errors, 54

Variables
    continuous, 2–3
    dependent, 101
    discrete, 2–3
    generalizations, 101
    relationship between, 76–77
    problems, 84
    standardized, 19
Variance, 17–18, 28. *See also* analysis of variance
Variation, 16–17, 19, 91

Unbiased estimates, 46–47

Yates' correction, 73–74